Structure of a chromosome

Eyewitness SCIENCE

Bromine in a flask

Written by
TOM JACKSON

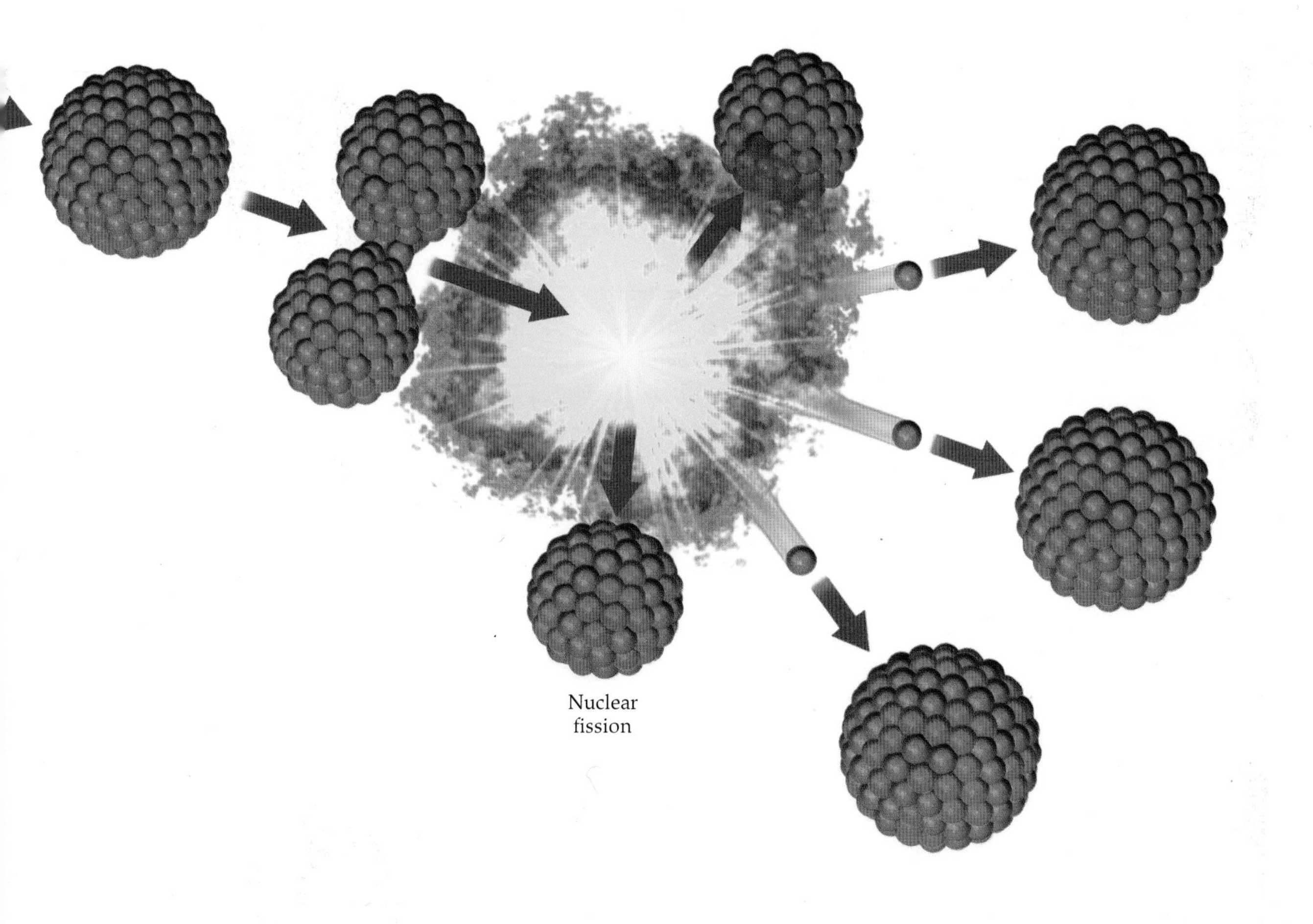
Nuclear fission

Solar probe

LONDON, NEW YORK, MELBOURNE, MUNICH, AND DELHI

Consultant Dr Donald R Franceschetti

DK DELHI

Project editor Ankush Saikia
Project designer Govind Mittal
Senior editor Kingshuk Ghoshal
Assistant editor Virien Chopra
Senior designer Mitun Banerjee
Designers Devika Dwarkadas, Shruti Soharia Singh
DTP designers Tarun Sharma, Jagtar Singh
Editorial manager Suchismita Banerjee
Design manager Romi Chakraborty
Production manager Pankaj Sharma
Head of publishing Aparna Sharma

DK LONDON

Senior editor Dr Rob Houston
Editor Jessamy Wood
Managing editor Julie Ferris
Managing art editor Owen Peyton Jones
Associate publisher Andrew Macintyre
Picture researchers Jo Walton, Karen VanRoss
Production editor Siu Yin Chan
Production controller Charlotte Oliver
Jacket designer Smiljka Surla

First published in Great Britain in 2011
by Dorling Kindersley Limited, 80 Strand, London WC2R 0RL

2 4 6 8 10 9 7 5 3 1

178350 - 07/11

A CIP catalogue record for this book is available from the British Library.

ISBN 978-1-4053-6206-1

Colour reproduction by MDP, UK
Printed and bound by
Toppan Printing Co. (Shenzhen) Ltd., China

www.dk.com

Bat using echolocation in flight

pH indicators

Lava lamp

Cinnabar, an ore of mercury

Zhang Heng's earthquake detector

Eyewitness
SCIENCE

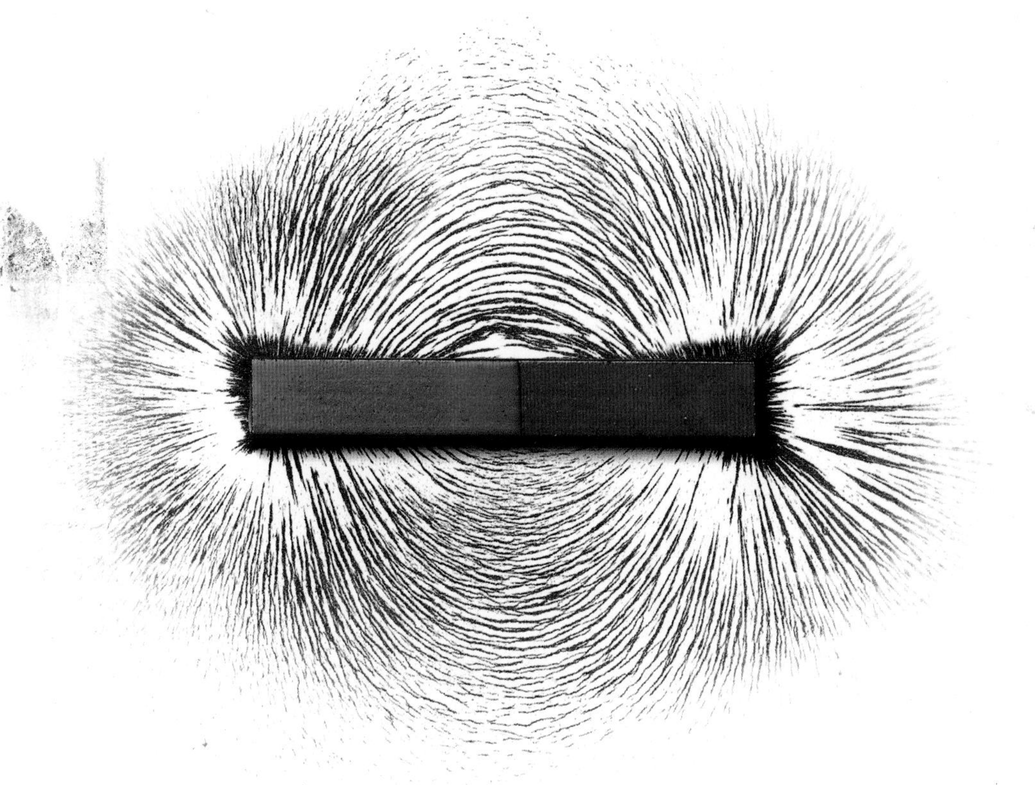

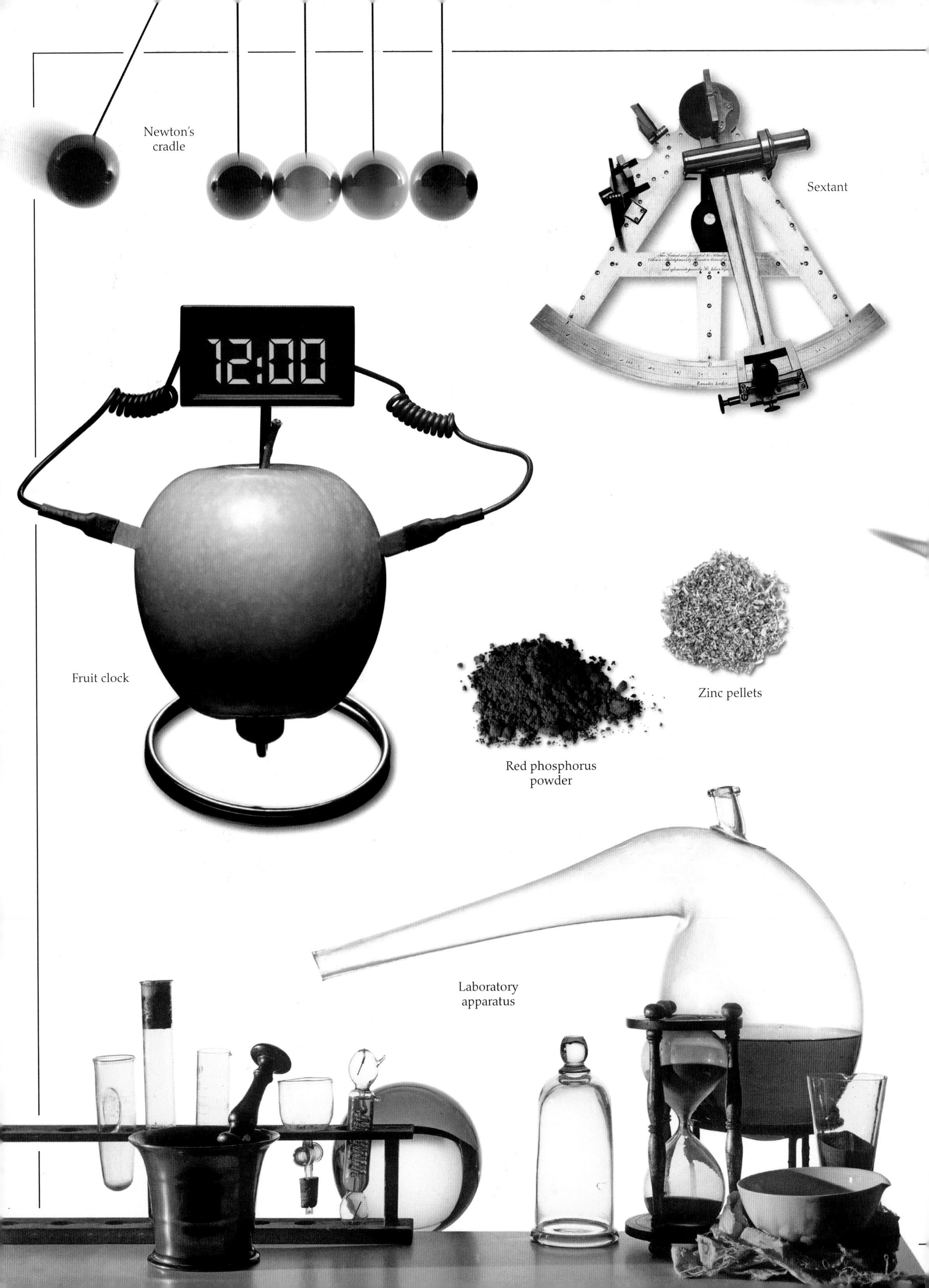

Newton's cradle

Sextant

Fruit clock

Zinc pellets

Red phosphorus powder

Laboratory apparatus

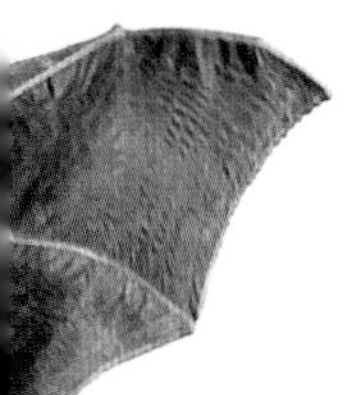

Contents

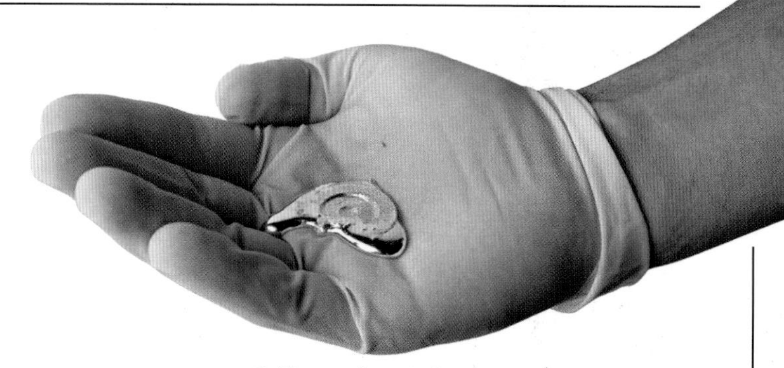

Gallium melting in hand

Making sense

THERE ARE MANY METHODS for finding answers to the mysteries of the Universe, and science is only one of these. However, science is unique. Instead of making guesses, scientists follow a system designed to prove if their ideas are true or false. They constantly re-examine and test their theories and conclusions. Old ideas are replaced when scientists find new information that they cannot explain. Once somebody makes a discovery, others review it carefully before using the information in their own research. This way of building new knowledge on older discoveries ensures that scientists correct their mistakes. Armed with scientific knowledge, people build tools and machines that transform the way we live, making our lives easier.

Fourteen lines of mathematical text in cuneiform script

ROUND NUMBERS
Scientists need to describe things exactly, and they do so using numbers and measurements. People have recorded their measurements for centuries. The Babylonians, who controlled parts of Iraq at times between 1900 and 600 BCE, recorded their numbers as marks in wet clay, which they then dried into solid tablets. We still use the ancient Babylonian counting system sometimes. They counted in 60s – not 10s. Today, seconds and minutes are added up in these Babylonian 60s, and we also divide circles into 360 degrees – 6 groups of 60.

SEEING PATTERNS
Scientists see a pattern in nature and try to explain it. Ancient people also tried to explain things in the natural world. The Mayans of Mexico observed the way days and nights get longer and shorter throughout the year. The equinoxes are the two days each year when day and night are the same length. The Mayans believed that on these special days, a snake goddess slithered down from the sky. They built a pyramid at Chichén Itzá to show this. When the Sun rises on an equinox, it casts a shadow on the pyramid's staircase. The steps make the shadow look like a slithering snake.

Snake-like shadow

Ibn al-Haythem appears on a stamp from the Arab state of Qatar

SEEING IS BELIEVING
Scientists should be good observers – but how do we know that what we see is coming from the outside world and not made up in our heads? This question was answered by the Arab scientist Ibn al-Haythem (or Alhazen) around 1,000 years ago. He was one of the first people to use experiments to prove ideas. He showed that light travels from objects into our eyes (see page 28), and not out of the eyes first, before bouncing back from the objects around us.

A physicist is dwarfed by a giant chamber that forms just a small part of the Large Haldron Collider

THE UNIVERSE AND EVERYTHING ELSE

Together, scientists have uncovered so much knowledge that no single person can remember it all. Different scientists are experts in different areas of science (see pages 68–69). This cave of pipes and wires is part of the Large Hadron Collider (see page 62), the largest machine ever made. It was built to investigate how the Universe was born billions of years ago. Scientists working here are physicists – the name comes from the Greek word *physis*, meaning nature. Physics is the study of matter, energy, and the forces inside everything in the Universe. Science can be used to investigate a lot more – mineralogists study crystals, meteorologists track changes in the weather, and malacologists are experts in slugs.

DIFFERENT VIEWS

Many scientists study the things found on Earth, but they do it in different ways. A biologist would be interested in the fact that this fossil was that of a trilobite (a marine organism) that lived 420 million years ago. A geologist would say that the rock was limestone, made from fragments of a shell that settled and fossilized on the ocean floor after the animal died. A chemist would be more interested in studying the composition of the rock. It contains calcium carbonate, which is a chemical made of carbon, calcium, and oxygen.

MAKING USE OF KNOWLEDGE

Pure science is driven by curiosity. Scientists solve puzzles about the Universe, from the way stars explode to how bumblebees fly. Applied science makes use of – or applies – what pure science has revealed to fix a problem. For example, this robotic arm has been built to work as much like a real one as possible. Physiologists have given it the same joints as a human arm while neurologists (nerve scientists) have worked out how to connect the robotic arm to a human's nerves so he can move it using his nerve impulses, just like a real arm.

Robotic limb copies movements of a real arm

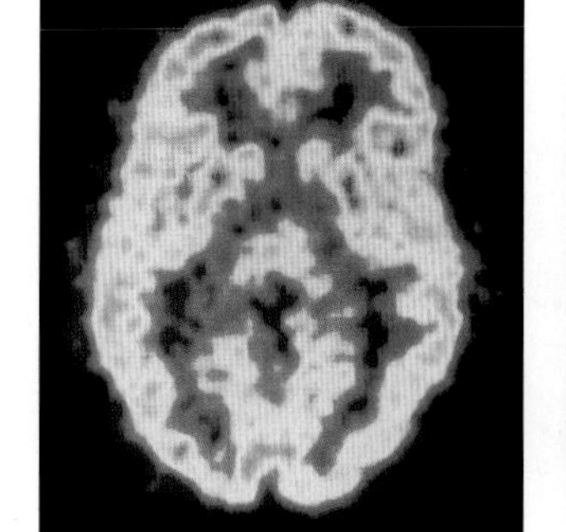

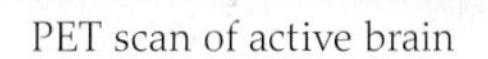

PET scan of active brain

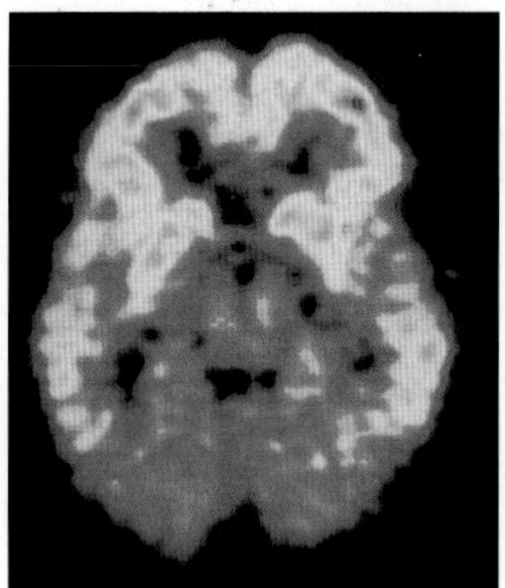

PET scan of resting brain

BEYOND SCIENCE

Scientists study things that they can record and measure. It is not yet possible to record what someone is thinking. One of the great mysteries of modern science is how the human brain works. Scientists can look inside a living brain using a PET scanner and see which areas of the brain are most active at which times. They are learning more every day, but we still do not know if the human brain is clever enough to figure itself out!

The foundations of science

ANCIENT PEOPLE EXPLAINED THEIR WORLD using stories known as myths. The Bakuba people of central Africa thought the world was formed when a giant was sick. The ancient Romans believed that storms and earthquakes were caused when Neptune, the god of the sea, was angry. Then from about 2,500 years ago, Greek philosophers such as Thales of Miletus and Aristotle began to question the workings of the Universe based on what they could see around them. They were the first people in recorded history to think as scientists, building knowledge by observing natural phenomena. Thinkers made new discoveries and developed new theories in other parts of the ancient world too, including Egypt, India, and China. Although some of the ideas of these pioneers were later proved wrong, their revolutionary ways of thinking laid the foundations of modern science.

15th-century illustration showing the four Greek elements

FOUR ELEMENTS
Ancient thinkers believed that everything was made from a handful of elements – simple substances that could not be split further. In Asia, people thought that there were five or six elements, but in Greece, most philosophers believed there were only four. These were fire, air, earth, and water. Everything in the world was considered to be made from a mixture of these elements.

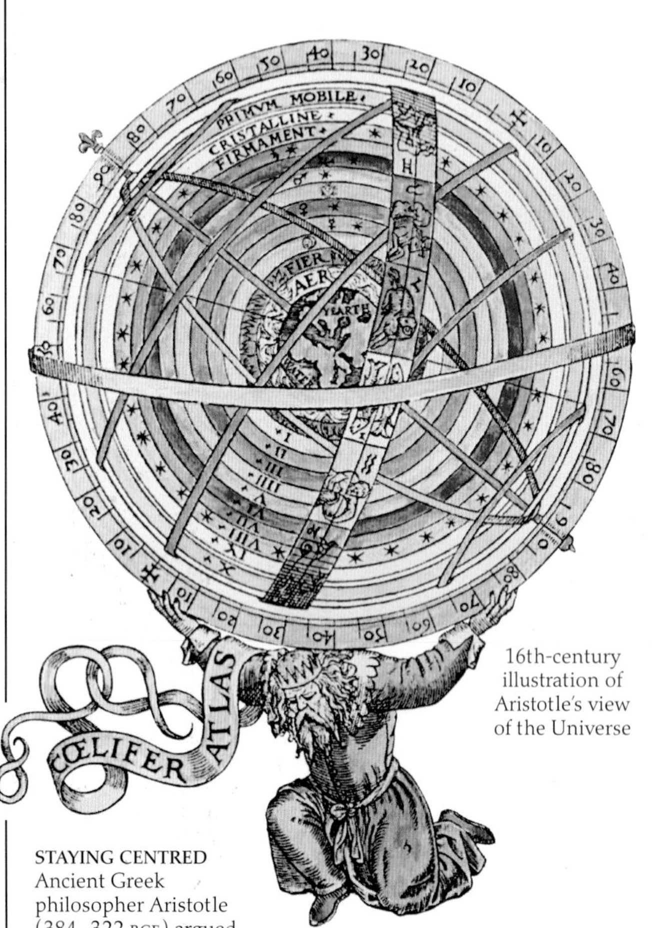

16th-century illustration of Aristotle's view of the Universe

STAYING CENTRED
Ancient Greek philosopher Aristotle (384–322 BCE) argued that the Universe was organized in rings. The element fire was at the centre of planet Earth, followed by bands of earth, water, and air. The Sun, Moon, and planets moved around Earth, while the stars formed the outermost ring. Aristotle had no proof, but his description fitted with what he could see and he assumed that Earth must be at the centre. His model of the Universe remained unchallenged for another 1,900 years.

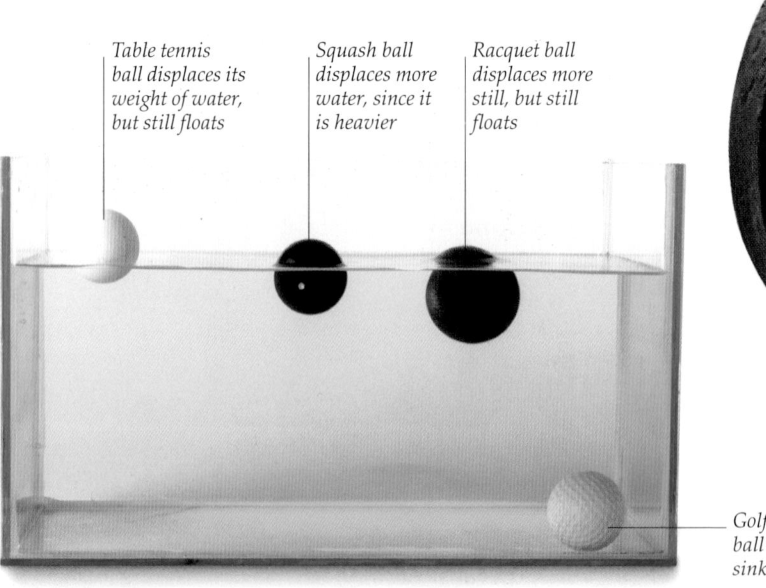

FLOATING IDEAS
The ideas of the Greek thinker Archimedes (287–212 BCE) are still in use today. One of these was about how things float. When Archimedes entered his bath, he realized that he displaced water. He worked out that if the weight of the displaced water is less than the weight of an object placed in water, then the object sinks. If not, it floats. This is why a heavy golf ball sinks in water while lighter balls float.

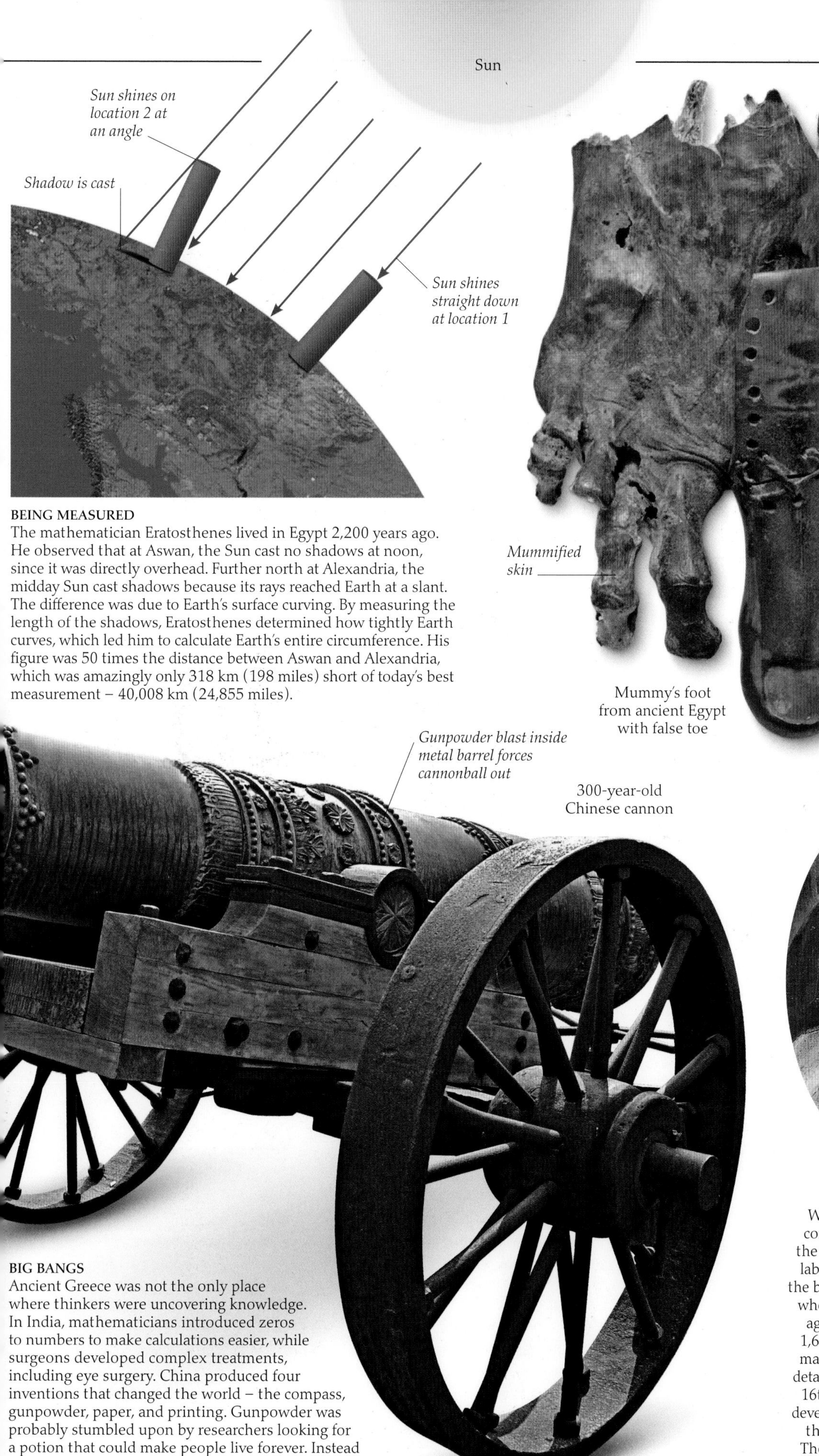

Mummy's foot from ancient Egypt with false toe

300-year-old Chinese cannon

HEALING POWERS

Ancient Egyptian doctors used their knowledge of medicine as much as their belief in magical rituals for healing. One of the first doctors in history known by name was Imhotep, who lived about 4,600 years ago in Egypt. Imhotep is thought to have written a manual with a very scientific description of the human body. An Egyptian doctor also made the first known false body part – a wooden big toe – some time around 3,000 years ago. Made for an Egyptian noblewoman, this wooden replacement allowed her to move around without any support.

BEING MEASURED

The mathematician Eratosthenes lived in Egypt 2,200 years ago. He observed that at Aswan, the Sun cast no shadows at noon, since it was directly overhead. Further north at Alexandria, the midday Sun cast shadows because its rays reached Earth at a slant. The difference was due to Earth's surface curving. By measuring the length of the shadows, Eratosthenes determined how tightly Earth curves, which led him to calculate Earth's entire circumference. His figure was 50 times the distance between Aswan and Alexandria, which was amazingly only 318 km (198 miles) short of today's best measurement – 40,008 km (24,855 miles).

FEMALE SCHOLARS

Women also made several important contributions to our understanding of the world. A waterbath used in modern laboratories to heat liquids is known as the bain-marie after Mary, a Jewish scholar who worked in Egypt about 2,000 years ago. Hypatia of Alexandria who lived 1,600 years ago, was a philosopher and mathematician. She can be seen in this detail from *The School of Athens*, a famous 16th-century painting. Hypatia helped develop the hydrometer – an instrument that measures the density of liquids. The hydrometer sinks deepest in thin, low-density liquids, such as petrol.

BIG BANGS

Ancient Greece was not the only place where thinkers were uncovering knowledge. In India, mathematicians introduced zeros to numbers to make calculations easier, while surgeons developed complex treatments, including eye surgery. China produced four inventions that changed the world – the compass, gunpowder, paper, and printing. Gunpowder was probably stumbled upon by researchers looking for a potion that could make people live forever. Instead they produced the first explosive, a powder that burns so fast and hot it creates a fireball.

Science at work

SCIENCE IS A POWERFUL TOOL. The principles discovered by science can be applied, or put to work – in fields such as medicine and engineering – to create newer and better technologies. Applied scientists develop technology – machines, tools, and techniques. However, technology has not always been scientific. Stone axes were a form of technology used by ancient people. Applying scientific knowledge has made modern technology a lot better. Instead of stone axes, engineers have created saws that cut with sharp diamonds, hot lasers, and even jets of water. Engineering applies science to machines. People invented simple machines in ancient times without fully understanding the science behind them. Later, scientists identified these simple machines – the lever, screw, wedge, ramp, pulley, and wheel – and recognized that complex machines work as combinations of them.

RAMPING UP

A ramp is a wedge working the other way around. Lifting this heavy piano onto the truck without a ramp would require a huge force to raise it quickly in one go. The ramp allows the force needed to do the work to be broken up into smaller sections. The workers use many small forces and not a large force, to push the piano slowly up in steps. A staircase also works like a ramp – with flat steps to make it easier to stand on.

THIN END OF A WEDGE

The first machine (and one of the earliest examples of technology) to be invented was the wedge. *Homo habilis,* an ancestor of *Homo sapiens* (modern human beings), chipped rocks into wedge-shaped cutters 2 million years ago. This early axe also functions as a wedge. It works when a downward force pushes on the wide end (the rear of the axe's head). The force is then focused onto the thin edge at the other end of the wedge (the axe's blade). As the same force is now squeezed into a smaller area, it pushes down strongly enough for the wedge to cut through whatever it is touching. Other wedge machines, such as metal knives, work in the same way.

TURNING THE SCREW

A screw can be seen as a ramp wrapped around a cylinder. As the screw is twisted into a plank, it drives into the wood. However, the screw's twisted shape stops it from falling out of the wood, making the screw perfect for holding things together. A screw has other purposes. The Archimedes screw – named after the ancient Greek philosopher – has been used to raise water for many centuries. The bottom of the screw is underwater. As the screw twists, the water gradually sloshes up the ramp.

MOVE ANYTHING
Levers work like off-centre seesaws. Pushing down at one end makes the other end rise. A small force applied at a distance from the turning point, or fulcrum, is multiplied many times by the lever, creating a strong force. Common levers include this claw hammer and a pair of scissors. Archimedes commented that if he had a lever long enough, and somewhere to stand, he could lift the world.

Small force applied far from fulcrum

Fulcrum

Force close to fulcrum is much greater, and hammer lifts nail

COG COMPUTER
The Antikythera mechanism is an ancient instrument found in a 2,100-year-old shipwreck near Crete. The machine once had dozens of connected wheels with teeth or cogs on the rim. The wheels would move in sequence, helping the device calculate the position of the Sun or Moon on any day of the year.

ROUND AND ABOUT
A wheel works like a disc of levers sticking out in all directions from the axle, which acts as the fulcrum. Each lever connects the axle with the wheel's outer edge – specifically the point where the tyre is touching the ground at any one time. The wheel is made to rotate by the boy forcing the pedals around. Lever action translates the small movements of the axle into large sweeps at the rim of the wheel. The tyre grips the ground, so as it turns, the wheel pushes backwards against the ground and the bike moves forward. The wheel is perhaps the most important simple machine. Wheeled carts and chariots were in use across the Middle East and eastern Europe about 6,000 years ago.

Axle forms the fulcrum around which the spokes rotate

Pulley

Bell

Blocks marking the hours of the day

19th-century gravity clock

Steel drum

Weight

KEEPING TIME
Clocks are machines that work at a fixed speed, so they count out time accurately. To do this they need to be driven by a constant, unchanging force. The simplest clocks make use of gravity, a force that pulls everything down to Earth. The oldest known clocks are stone water flasks used in Egypt 3,400 years ago. In these, gravity pulled a trickle of water though a hole in the flask and it always took the same time for the flask to empty out. This clock uses a pulley and relies on gravity pulling on a weight attached to a steel drum. This slowly moves down the blocks marking the hours. A system of levers rings the bell when the drum reaches the bottom, signalling that it is time to wind it back up.

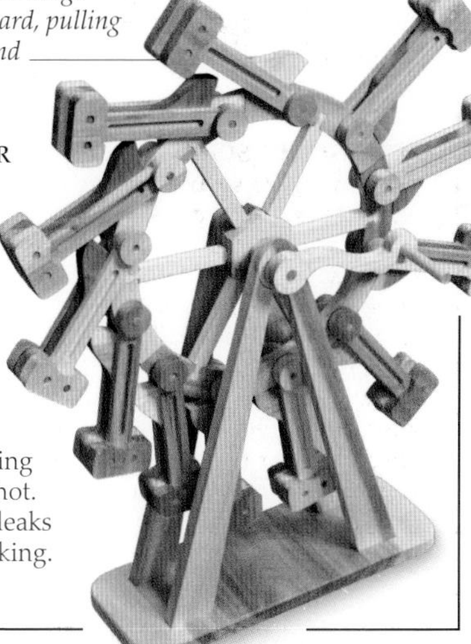

Hammer on a hinge moves forward, pulling wheel around

WORKING FOREVER
This wheel from 1235 CE was meant to keep spinning forever. The idea was that the top weight would flop forward, pulling the wheel around until the next weight could tip over, and so on. But, like all machines, it would stop moving unless given a push from time to time. It was only in the 19th century that scientists realized why perpetual motion is impossible. The moving parts of a machine rub together and get hot. Gradually, all the energy in the system leaks away as heat, and the machine stops working.

Changing the world

Orrery showing planets revolving around the Sun

Sometimes, scientists discover that what they once thought to be true is in fact wrong. In 1543, Nicolaus Copernicus announced his discovery that would turn people's view of the world upside down. Until that time, people believed that Earth was at the centre of the Universe, and everything else – including the Sun, planets, and stars – moved around it. But some planets slowed or sped up across the sky, and even went into reverse. Copernicus showed that he could explain this behaviour if Earth was just one of several planets moving around the Sun. Copernicus's discovery challenged facts that people had always accepted. There were more challenges to come. During the next decades, scholars in Europe looked again at established knowledge, especially from ancient Greek thinkers, and began a revolution in science.

DANGEROUS IDEA
In the 16th century, the Catholic Church controlled much of Europe and insisted that Earth was at the centre of the Universe. Anyone contradicting the Church's belief – or even making a model, or orrery, of the planets moving around the Sun – could be jailed! Copernicus knew the risks and kept quiet about his discovery until just before he died.

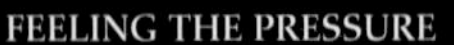

FEELING THE PRESSURE
Ancient Greeks believed that the *Anemoi* – wind gods – controlled the seasons and the weather. In 1643, the Italian scientist Evangelista Torricelli became the first person to use science to forecast the weather. He built a barometer – a U-shaped column filled with the liquid metal mercury. It measured air pressure – the force of the atmosphere pushing at Earth's surface. The air pushed at the open end of Torricelli's tube and made the mercury rise on the other side. The barometer showed that when air pressure was high, the day would probably be dry. Today's weather maps connect points of equal air pressure with lines called isobars. These help predict how the weather will change.

De Humani Corporis Fabrica by Andreas Vesalius

THE DEAD TELL NO LIES
The Roman physician Galen was the established authority on human anatomy – the study of body structure – for centuries. In the 16th century, the Belgian physician Andreas Vesalius began dissecting human bodies to examine anatomy. He made highly detailed drawings of the body, inside and out. Vesalius published his work in a book known as *De Humani Corporis Fabrica* (Fabric of the Human Body). His pictures were the first accurate record of human anatomy and corrected errors in much of Galen's work.

Isobar shows places with equal pressure

Replica of Torricelli's barometer

Weather map showing circular regions of high and low pressure

GETTING EVEN
Greek thinker Aristotle had said that heavy things fall to the ground faster than light objects. However, in the 16th century, a young scientist called Galileo Galilei proved Aristotle wrong. Galileo dropped two cannonballs – one large and one small – from the Leaning Tower of Pisa, Italy. He noted that the balls fell evenly and hit the ground at the same time. He realized that objects fall to Earth in the same time, whether they are heavy or not.

Near side of the Moon

Tycho crater, one of the Moon's features picked out by Galileo in his drawings

A terrella

POINTING NORTH

For centuries it was well known that a swinging magnet always pointed in the same direction – north. In 1600, William Gilbert explained why. Any small magnet is attracted to a much larger one – Earth itself. He proved this idea by making a terrella, a model of Earth, out of a spherical magnet. By moving a smaller magnet around the surface of the terrella, Gilbert showed that the little magnet always swung towards the terrella's north pole. Gilbert suggested that under Earth's rocky crust, the planet was made up of a giant ball of iron, which produced the magnetism (see page 42).

SEEING FURTHER

About 400 years ago, merchants in Venice, Italy, used telescopes to watch for their ships approaching the city. Galileo, who lived in Venice at that time, built his own powerful telescope and used it to magnify the night sky. He was amazed to see that the Moon was covered with mountains and craters, like a smaller version of Earth. He studied these features and made detailed drawings. Galileo also found that Jupiter was orbited by not one, but four moons. This was the first direct evidence to support Copernicus's research – not everything in the Universe revolved around Earth.

From magic to method

THE WORD SCIENTIST WAS INVENTED just 180 years ago. Until then, researchers were known as natural philosophers or alchemists. These people did not use modern scientific methods. Most alchemists believed in magic, and many of them searched for potions giving superhuman powers. Alchemists had their uses. They discovered new dyes and ways to make valuable perfumes. As they searched for other useful substances, alchemists began to keep records so they could repeat procedures. They also used experiments to test ideas. Slowly, alchemists were beginning to work like scientists do today.

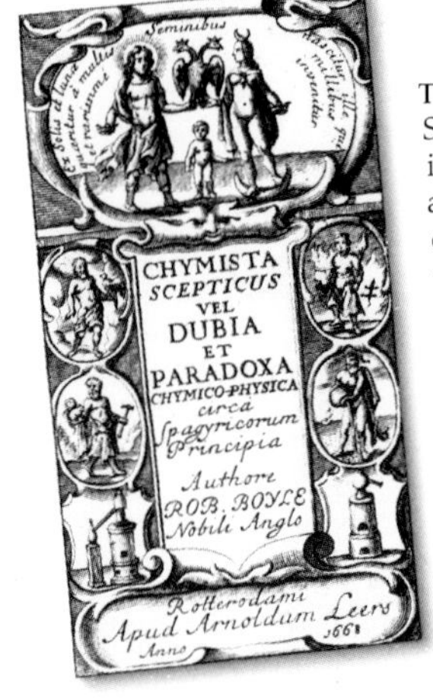
CHYMISTA
SCEPTICUS
VEL
DUBIA
ET
PARADOXA
CHYMICO-PHYSICA
circa
Spagyricorum
Principia
Authore
ROB. BOYLE
Nobili Anglo
Roterodami
Apud Arnoldum Leers
Anno 1668

THE LAST ALCHEMISTS
Some of the greatest figures in the history of science were alchemists. Isaac Newton, who explained how gravity works, was an alchemist looking for magical ways to turn lead into gold. Newton's friend Robert Boyle did similar research, but in 1661, he wrote a book called *The Skeptical Chymist* – about how the behaviour of metals, gases, and crystals should be explained using science instead of magic.

A 19th-century painting called *The Alchemist*

HOCUS POCUS
Alchemists were the first people to attempt to explore what the world was made of. Most of them searched for the philosopher's stone – a substance believed to turn common metals into precious ones – or the elixir of life, which could supposedly make people immortal. Alchemists kept their work secret to prevent others from stealing their ideas. An 8th-century Arab alchemist, Geber, recorded his results in a code that was meaningless to others. The word gibberish comes from his name.

IN THE LAB

The apparatus used in an alchemist's workshop would be familiar to a modern chemist. Bell jars collected the gases released, while any liquids produced in the experiments dripped from the spout of a retort. Hot gases were made into liquids by running them through a cooling condenser. Alchemists stored liquids in test tubes and retorts and used burettes to measure liquids to the exact drop. Simple clocks, such as a sand timer, made sure that mixtures boiled for the right length of time.

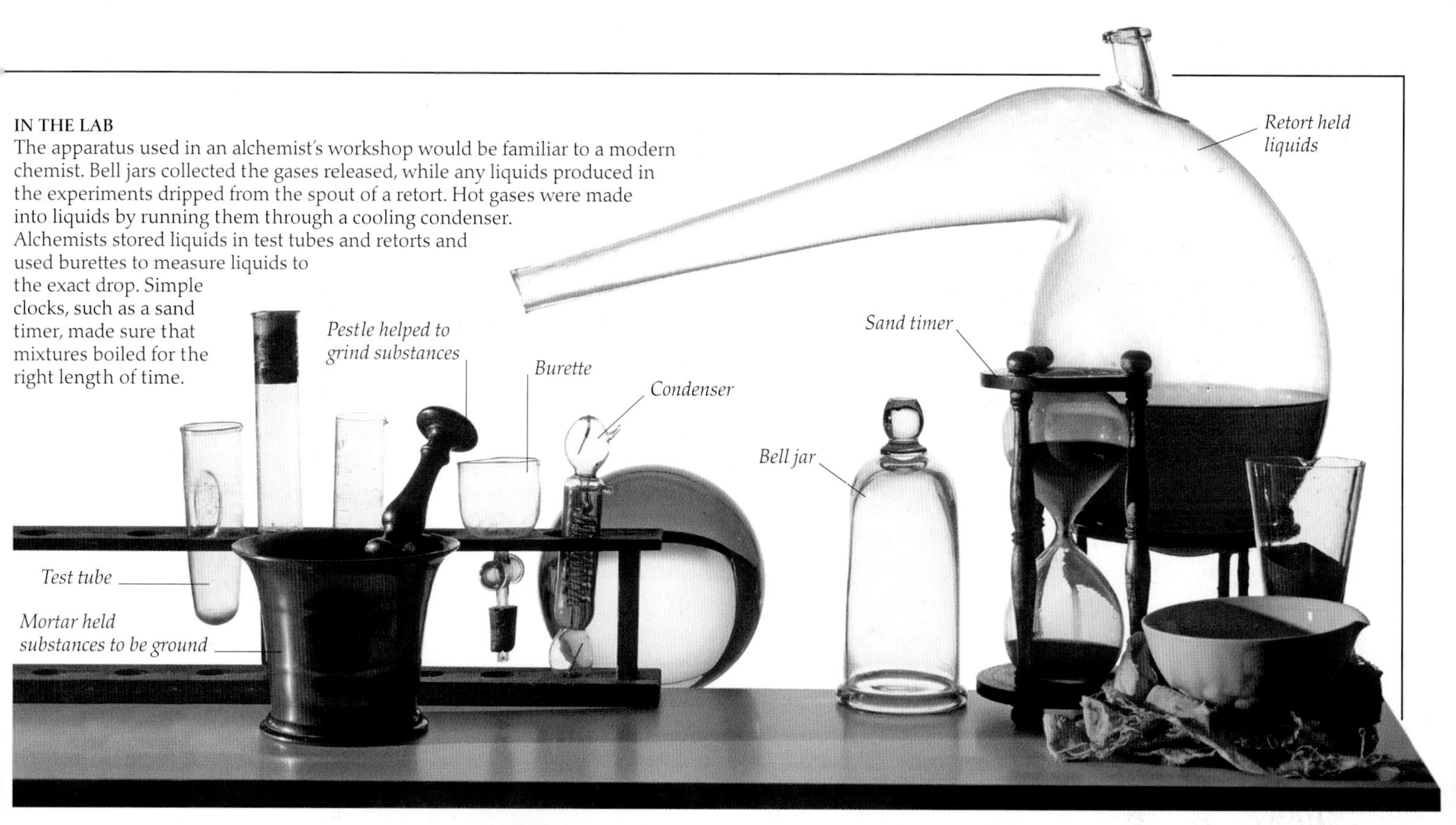

Retort held liquids

Pestle helped to grind substances

Burette

Condenser

Sand timer

Bell jar

Test tube

Mortar held substances to be ground

A NEW WAY

In 1620, the English philosopher Francis Bacon set out a new method for research in a book called *Novum Organum*. He believed that thinkers should question the natural world and follow a specific method that included observation, suggestion of a theory, and experimentation. He stressed the sharing of information using simple language so that more people could understand and learn. This is the basis of the modern scientific method.

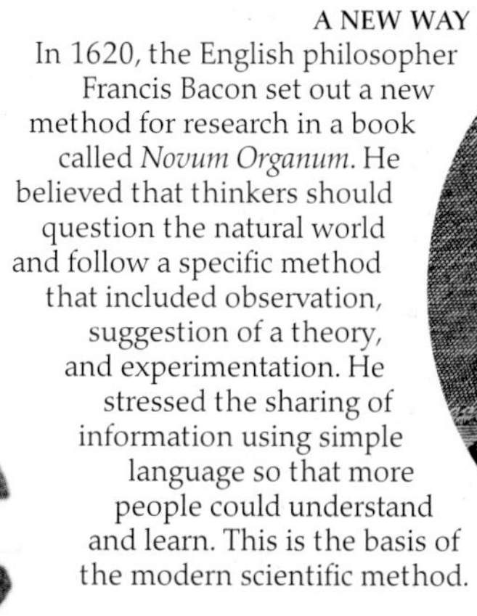

BEATING OF THE HEART

Doctors once thought that blood was made continuously by the liver and heart before being pumped into the body, where it was used up. In 1628, the English doctor William Harvey carried out experiments that showed that if this were true, a person would have to make 250 kg (550 lb) of blood a day! He showed that the body had a fixed amount of blood, which circulated away from the heart in arteries and then returned in veins.

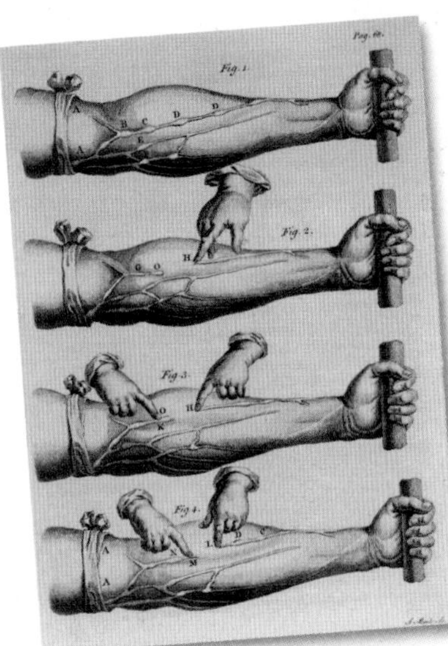

Harvey's drawings show valves in the veins that stop blood from flowing backwards

LOOKING CLOSER

People are limited by what their eyes can see. To see more, scientists invented machines to detect things that are too small or too far away. In 1665, Englishman Robert Hooke used a simple microscope to take a closer look at insects and plants. He found a tiny world, full of detail. In the 1670s, the Dutch lensmaker Antonie van Leeuwenhoek made a better microscope. Looking through it, he became the first person to see bacteria, which he named "animalcules".

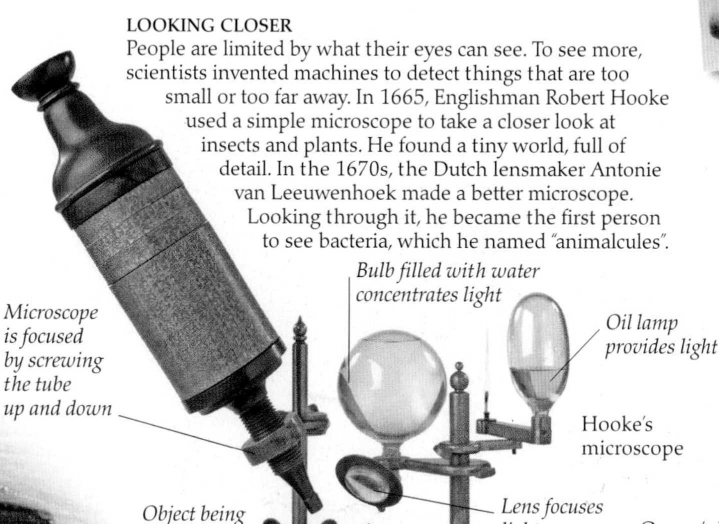

Bulb filled with water concentrates light

Oil lamp provides light

Microscope is focused by screwing the tube up and down

Hooke's microscope

Object being studied

Lens focuses light

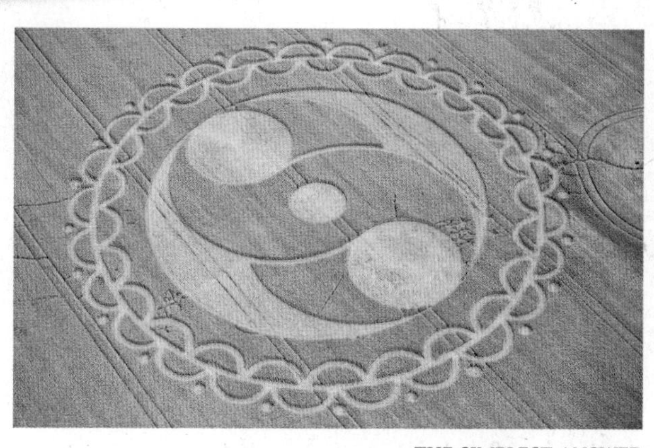

THE SIMPLEST ANSWER

Occam's Razor is a rule that helps scientists form ideas. Named after 14th-century English thinker William of Ockham, it says that the simplest answer is most likely the correct one. When mysterious patterns of flattened crops appeared in fields in the 1970s and 1980s, some suspected that these "crop circles" were made by freak weather or even alien spacecraft. Using Occam's Razor, however, the best explanation is that they were created by people as practical jokes. This turned out to be true.

A new order

Ball in motion hits row of balls with a force

FEW SCIENTISTS MAKE A DISCOVERY that fundamentally changes science, but Isaac Newton did just that. In the 1680s, he produced a series of laws that explained how things moved. Newton's most famous achievement was the discovery of how gravity worked. Gravity is a force that pulls masses together. The pull of larger bodies, such as Earth, is much stronger than that of smaller ones, such as apples. Newton described the force using mathematical calculations, which showed that a falling apple and an orbiting moon both move according to the same set of rules. Newton's laws of motion provided a new way to understand and investigate the Universe.

DOWN TO EARTH
Newton is said to have figured out the way gravity works after seeing an apple fall from a tree. He suggested that every object has a pull of gravity. While Earth pulls on an apple, the apple also pulls on Earth. Because Earth is much heavier, its gravity is also stronger, so the apple accelerates towards Earth, while Earth hardly moves at all.

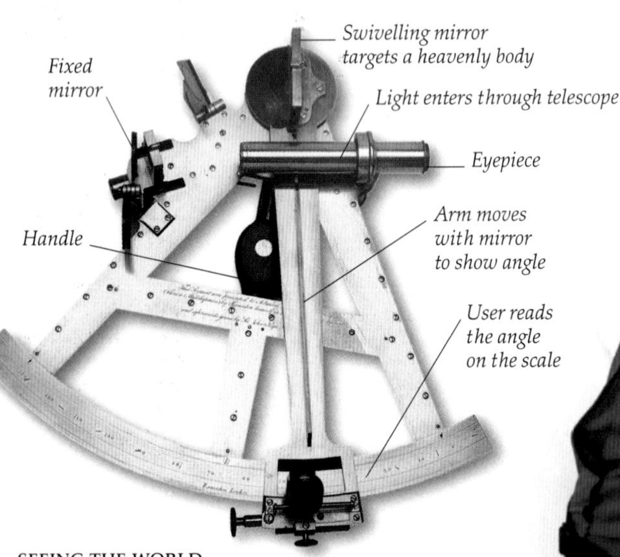

SEEING THE WORLD
Sailors once used a sextant to measure the angle of the Sun, Moon, or a star above the horizon. This knowledge, combined with a ship's almanac (a set of tables), allowed the sailor to calculate the ship's latitude – its position north or south of the Equator. From the 1770s, ships have carried almanacs, which predict the time when certain heavenly bodies will be at their highest point in the sky that day. This is thanks to Newton and other scientists showing how the objects move in space.

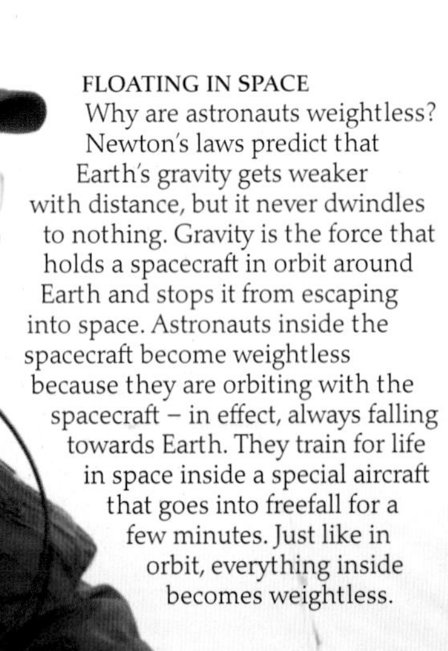

FLOATING IN SPACE
Why are astronauts weightless? Newton's laws predict that Earth's gravity gets weaker with distance, but it never dwindles to nothing. Gravity is the force that holds a spacecraft in orbit around Earth and stops it from escaping into space. Astronauts inside the spacecraft become weightless because they are orbiting with the spacecraft – in effect, always falling towards Earth. They train for life in space inside a special aircraft that goes into freefall for a few minutes. Just like in orbit, everything inside becomes weightless.

THE SPEED OF LIGHT
In 1676, the Danish astronomer Ole Rømer measured the speed of light using Io, a moon of the planet Jupiter. Newton's laws of motion predicted Io's motion. It is hidden when it moves behind Jupiter, but Rømer knew the exact times Io would reappear. However, he found that the moon appeared 10 minutes late – the time its light took to reach Earth. The delays got longer as Earth orbited away from Jupiter. Rømer calculated that light traveled at 220,000 km (137,000 miles) per second – only 25 per cent slower than the true value.

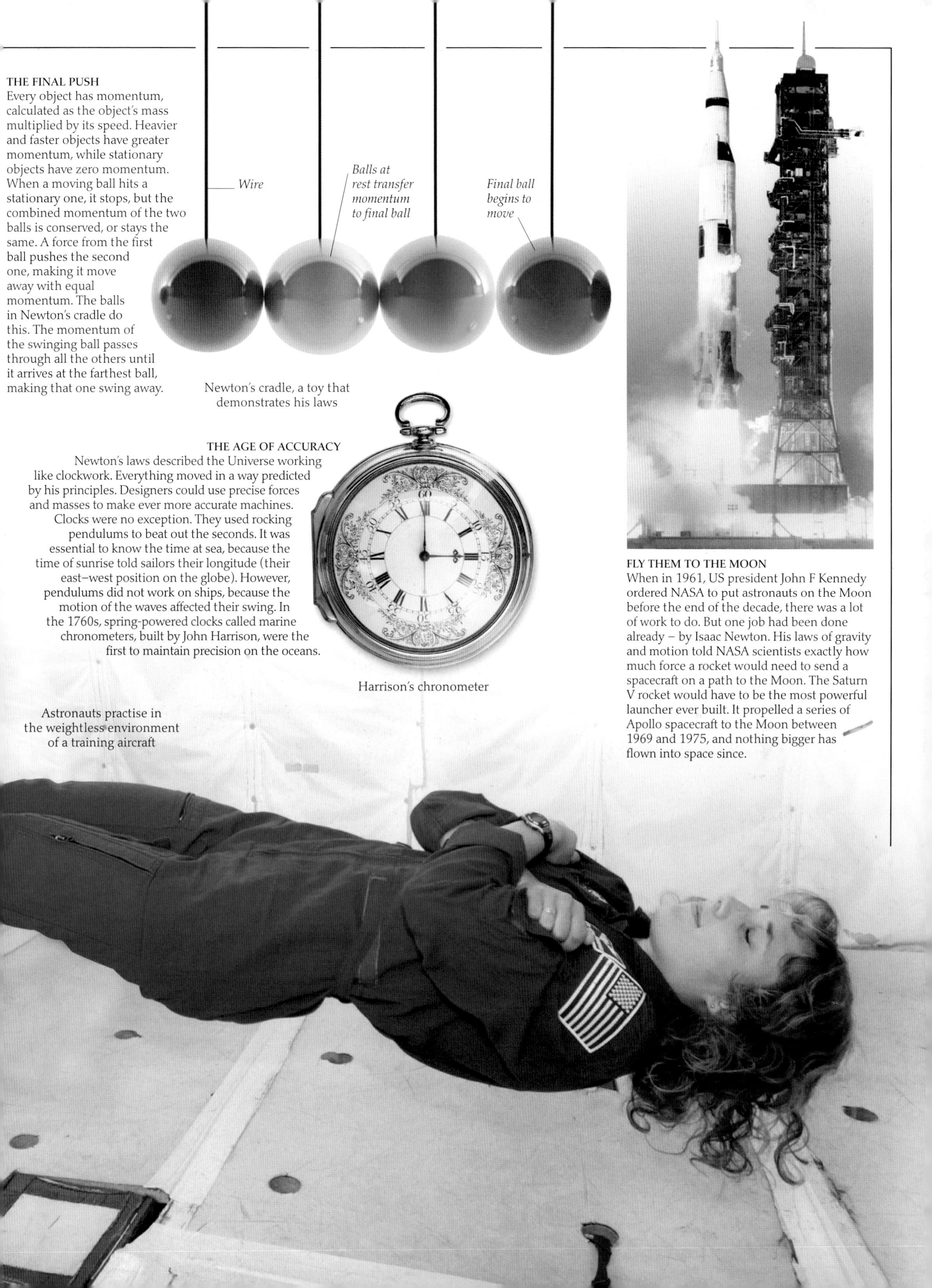

THE FINAL PUSH
Every object has momentum, calculated as the object's mass multiplied by its speed. Heavier and faster objects have greater momentum, while stationary objects have zero momentum. When a moving ball hits a stationary one, it stops, but the combined momentum of the two balls is conserved, or stays the same. A force from the first ball pushes the second one, making it move away with equal momentum. The balls in Newton's cradle do this. The momentum of the swinging ball passes through all the others until it arrives at the farthest ball, making that one swing away.

Newton's cradle, a toy that demonstrates his laws

THE AGE OF ACCURACY
Newton's laws described the Universe working like clockwork. Everything moved in a way predicted by his principles. Designers could use precise forces and masses to make ever more accurate machines. Clocks were no exception. They used rocking pendulums to beat out the seconds. It was essential to know the time at sea, because the time of sunrise told sailors their longitude (their east–west position on the globe). However, pendulums did not work on ships, because the motion of the waves affected their swing. In the 1760s, spring-powered clocks called marine chronometers, built by John Harrison, were the first to maintain precision on the oceans.

Harrison's chronometer

FLY THEM TO THE MOON
When in 1961, US president John F Kennedy ordered NASA to put astronauts on the Moon before the end of the decade, there was a lot of work to do. But one job had been done already – by Isaac Newton. His laws of gravity and motion told NASA scientists exactly how much force a rocket would need to send a spacecraft on a path to the Moon. The Saturn V rocket would have to be the most powerful launcher ever built. It propelled a series of Apollo spacecraft to the Moon between 1969 and 1975, and nothing bigger has flown into space since.

Astronauts practise in the weightless environment of a training aircraft

Scientific method

MODERN SCIENTISTS WORK BY following the scientific method. They begin by asking a question. Next, they find out more about their problem by collecting information, which helps them to come up with a possible answer to the question. This idea is known as a hypothesis, which is then tested with experiments. They design experiments that may or may not produce results predicted by the hypothesis. An unexpected result means that the hypothesis must be wrong. Scientists publish their results to let others know what they have found. Other scientists repeat the experiment to ensure that the original experiment was performed correctly.

QUESTION
Scientists question everything. For example, why is the tongue of a blue-tongued skink blue? Very few animals in nature have this colouring, and the rest of this chunky Australian lizard is coloured to blend in with its desert surroundings. Biologists have learnt that the skink uses the vibrant colour of its tongue to frighten away its enemies.

IDEA
The best scientific ideas have the power to predict. In 1974, two scientists were studying CFCs, chemicals used inside spray cans and refrigerators, when they discovered that CFCs reacted with a type of oxygen called ozone. This gave them the idea, or hypothesis, that CFCs might destroy the protective layer of ozone high in Earth's atmosphere. In 1985, an enormous hole was found in the ozone layer, which supported their hypothesis. CFCs are now banned, and the ozone layer is regenerating.

Camera looks at the area around the rover

Radio antenna sends information to Earth

Solar panels collect energy from sunlight to power the rover

Rover analysing soil samples on Mars

OBSERVATION
Investigation helps scientists form better ideas. Just 100 years ago, some scientists thought that the planet Mars was covered with canals. They claimed that this was evidence of an ancient Martian civilization. It was only when the first spacecraft landed on the planet in 1976, that they found out that Mars was a lifeless desert. However, later probes sent to Mars found evidence that simple alien life might have existed there billions of years ago. The latest NASA rovers on Mars are collecting more information to test this idea.

Pendulum swings freely in one vertical plane above the dial

EXPERIMENT
Scientists have known for centuries that Earth spins around every 24 hours. That is how they explain why the Sun rises and sets each day. However, no experiment directly demonstrated Earth's spin until 1851, when physicist Léon Foucault set up this giant pendulum in Paris. The pendulum swings in a fixed plane – it just sweeps back and forth and never swings in other directions. However, after swinging for hours above the dial, it seemed as if Foucault's pendulum was moving clockwise. It was actually the dial that had rotated anticlockwise beneath it. The dial was moving as Earth rotated, while the pendulum kept swinging in its fixed plane. At last, here was evidence that the world turned.

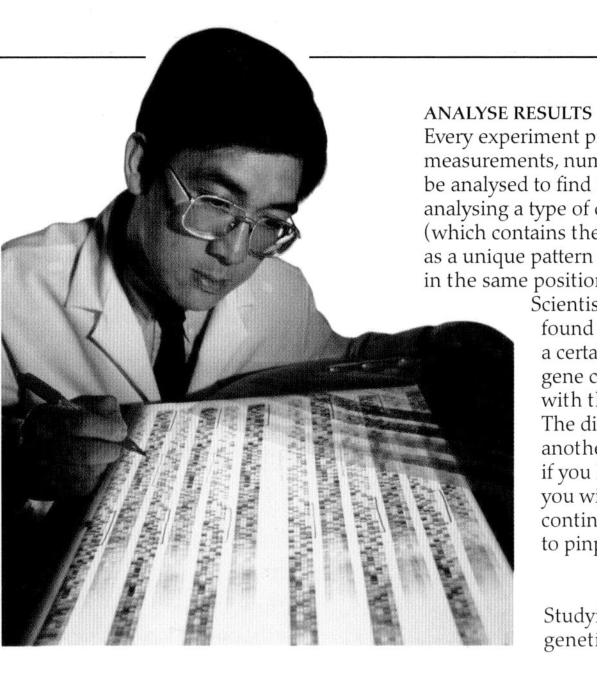

ANALYSE RESULTS

Every experiment produces data – a set of measurements, numbers, or patterns – that has to be analysed to find out what it means. This scientist is analysing a type of data that shows a person's DNA (which contains the person's genes, see page 50) as a unique pattern of stripes. People with a stripe in the same position might possess the same gene. Scientists studying data like this have found that most people suffering from a certain disease of the skeleton share a gene called B27. However, most people with this gene do not get the disease. The disease could actually be caused by another gene sitting close to B27 – so if you have the disease-causing gene, you will have B27 as well. Researchers continue to analyse genetic patterns to pinpoint this faulty gene.

Studying genetic data

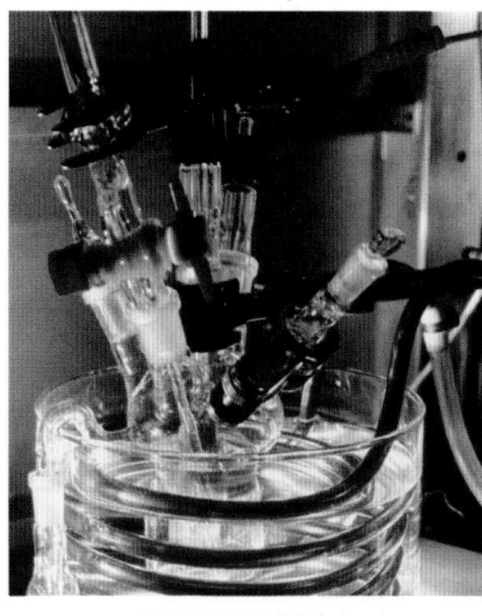

REPEATABILITY

Scientific experiments must be repeatable so that the results can be verified. In 1989, two scientists claimed that they had discovered "cold fusion" – a method of generating nuclear power using the apparatus seen here. However, when other scientists conducted the experiment, they found that the system did not work. The claims of cold fusion had been untrue.

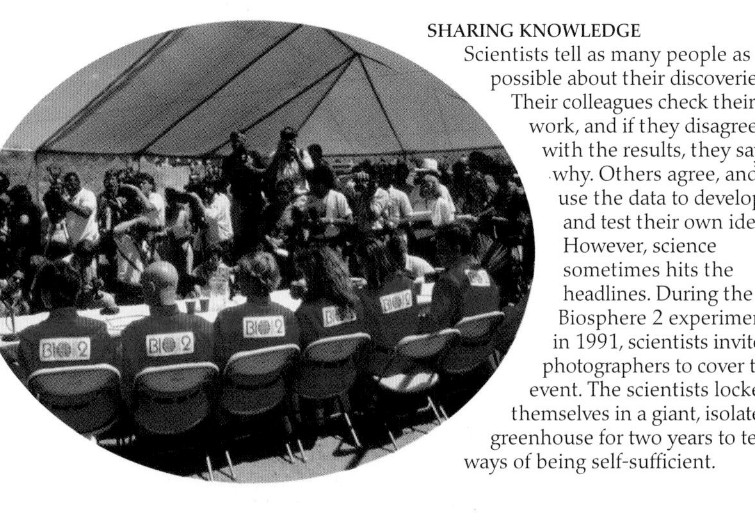

SHARING KNOWLEDGE

Scientists tell as many people as possible about their discoveries. Their colleagues check their work, and if they disagree with the results, they say why. Others agree, and use the data to develop and test their own ideas. However, science sometimes hits the headlines. During the Biosphere 2 experiment in 1991, scientists invited photographers to cover the event. The scientists locked themselves in a giant, isolated greenhouse for two years to test ways of being self-sufficient.

INSPIRATION

Scientists are often inspired by discoveries made by others. In 2010, the biologist Craig Venter decided that people had such an understanding of how genes control living organisms that he could build an artificial life form. He chose a tiny life form called a bacterium and removed all of its natural genes. He put together a series of artificial genes and placed them inside the bacterium. The modified bacterium lived and reproduced normally. It is similar to the *Mycoplasma* bacteria seen here.

Markers help to measure the angle by which the plane of the pendulum's swing rotates

Dial rotates in an anticlockwise direction in the Northern Hemisphere

Mercury, or quicksilver, is the only liquid metal at room temperature

Many elements

EVERYTHING IN THE UNIVERSE is made up of combinations of different elements. An element is a substance that cannot be divided any further into simpler ingredients. Some elements have been familiar for centuries, such as iron and copper, used by ancient metalworkers, but no one identified them as elements until the 17th century. The first recorded discovery of an element was by Hennig Brandt, a German alchemist. In the 1660s, he discovered phosphorus after boiling his urine for hours. A white powder was left behind and Brandt was amazed to see that it glowed in the dark. He named the substance phosphorus after the Greek word for "light-giver".

Copper was one of the first elements to be purified, about 9,000 years ago

Phosphorus exists in red and white coloured forms

METALS

Three-quarters of the Universe's 90 or so elements are metals. They are hard, heavy solids, which often shine when polished. Pieces of metal do not break easily, but can be hammered and bent into many shapes, which makes them very useful substances. Some metals, such as gold, are found in a pure form in nature. They stay pure even after hundreds of years, which is one of the reasons why gold is so precious today.

Whitish zinc is mixed with reddish copper to make golden brass

DIFFERENT ELEMENTS

There are about 90 elements that occur naturally on Earth, and several more that have been created in laboratories. Each element has its own set of unique characteristics. Most elements on Earth are solid, while the others are mainly gases. Mercury and bromine are the only two elements that are liquids at room temperature.

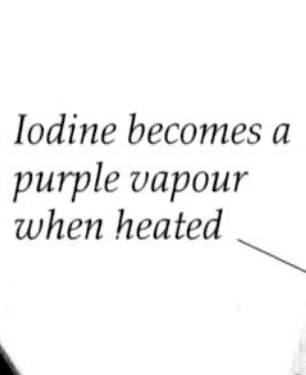

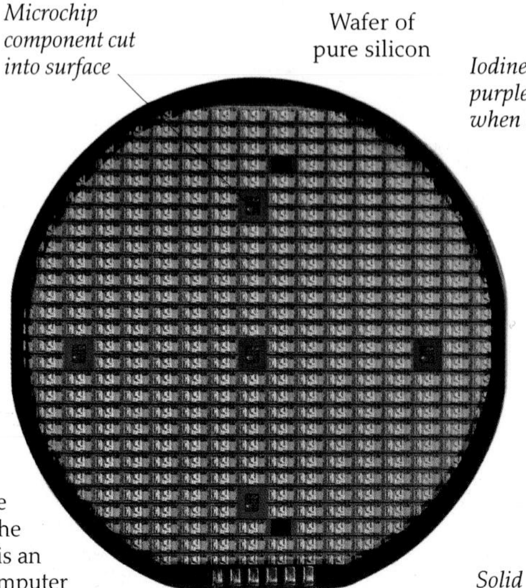

Microchip component cut into surface

Wafer of pure silicon

SEMIMETALS

Seven elements are semimetals – their properties are halfway between those of metals and non-metals. Semimetals are solid and shiny, like metals, but they are also quite soft and likely to crumble, like non-metals. The most common semimetal is silicon, which is an ingredient in sand and many rocks. Computer systems are run by microchips made of silicon mixed with a few other elements, such as aluminium.

Iodine becomes a purple vapour when heated

Solid iodine

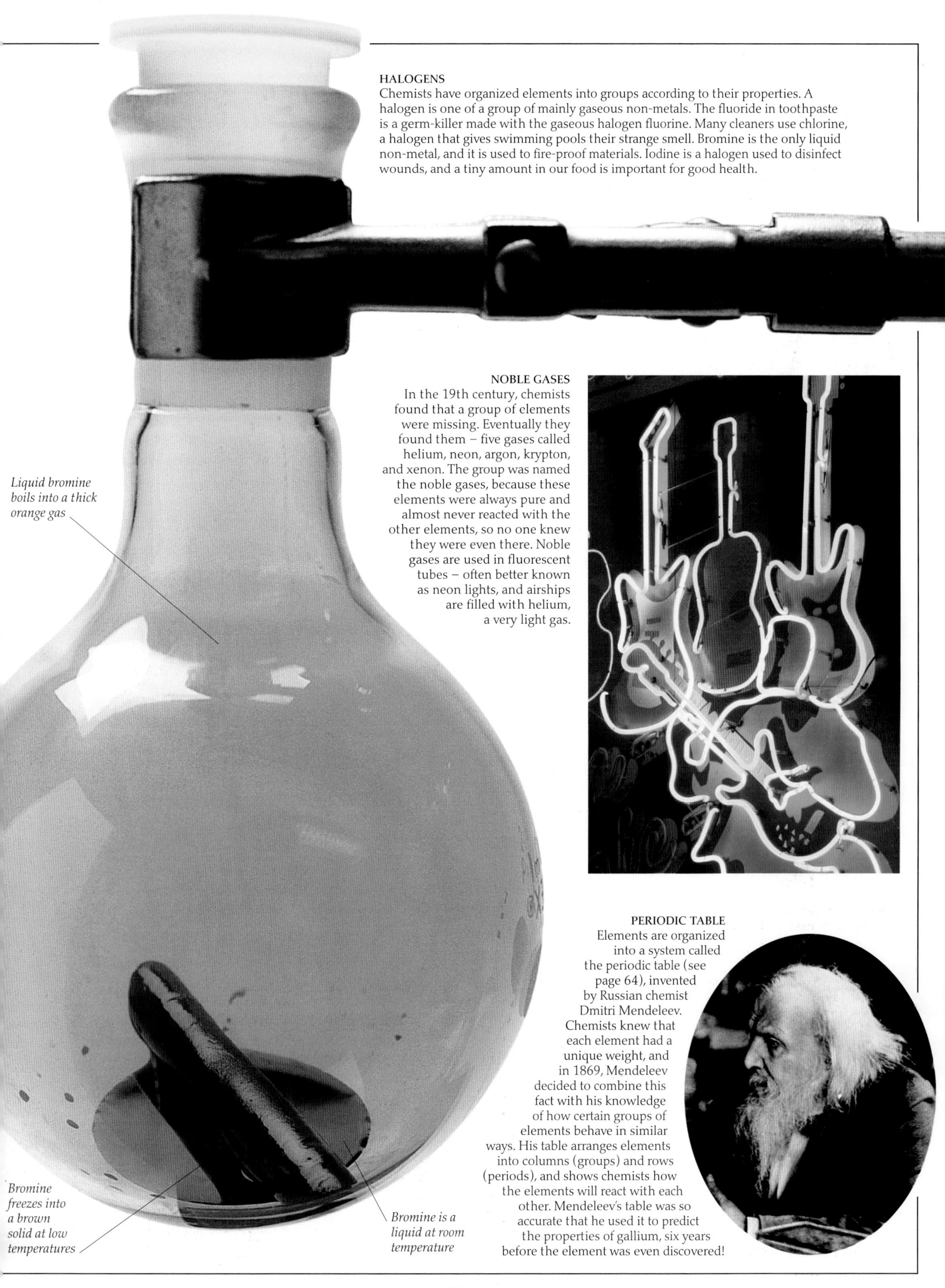

HALOGENS

Chemists have organized elements into groups according to their properties. A halogen is one of a group of mainly gaseous non-metals. The fluoride in toothpaste is a germ-killer made with the gaseous halogen fluorine. Many cleaners use chlorine, a halogen that gives swimming pools their strange smell. Bromine is the only liquid non-metal, and it is used to fire-proof materials. Iodine is a halogen used to disinfect wounds, and a tiny amount in our food is important for good health.

NOBLE GASES

In the 19th century, chemists found that a group of elements were missing. Eventually they found them – five gases called helium, neon, argon, krypton, and xenon. The group was named the noble gases, because these elements were always pure and almost never reacted with the other elements, so no one knew they were even there. Noble gases are used in fluorescent tubes – often better known as neon lights, and airships are filled with helium, a very light gas.

PERIODIC TABLE

Elements are organized into a system called the periodic table (see page 64), invented by Russian chemist Dmitri Mendeleev. Chemists knew that each element had a unique weight, and in 1869, Mendeleev decided to combine this fact with his knowledge of how certain groups of elements behave in similar ways. His table arranges elements into columns (groups) and rows (periods), and shows chemists how the elements will react with each other. Mendeleev's table was so accurate that he used it to predict the properties of gallium, six years before the element was even discovered!

Continued on next page

Continued from previous page

BURNING GAS

Hydrogen is the most common element in the Universe and makes up three-quarters of all matter. When astronomers look out into space, most of what they see is hydrogen burning in huge balls of fire – the stars. New stars form from gigantic clouds of hydrogen in space. Hydrogen is also found on Earth. It combines with oxygen to make water. The name hydrogen means "water maker".

Pillar-like masses of hydrogen gas in the Eagle Nebula

Climber breathes oxygen through mask as oxygen is scarce at high altitudes

Oxygen supply carried in a tank in rucksack

ELEMENT OF LIFE

All animals and plants need a supply of oxygen to survive. Oxygen is a gas and humans and other land animals get their oxygen by breathing it in from the air. Pure oxygen makes up only about one-fifth of the air – the rest mainly being nitrogen – but this is normally enough to supply the body. However, climbers in mountains face thin air and must take their own supply of oxygen with them. A lot more oxygen is locked away as an ingredient in Earth's water and rocks. It alone makes up nearly half of the weight of Earth's crust. Only iron is more common on Earth and most of that is deep inside Earth's metal core.

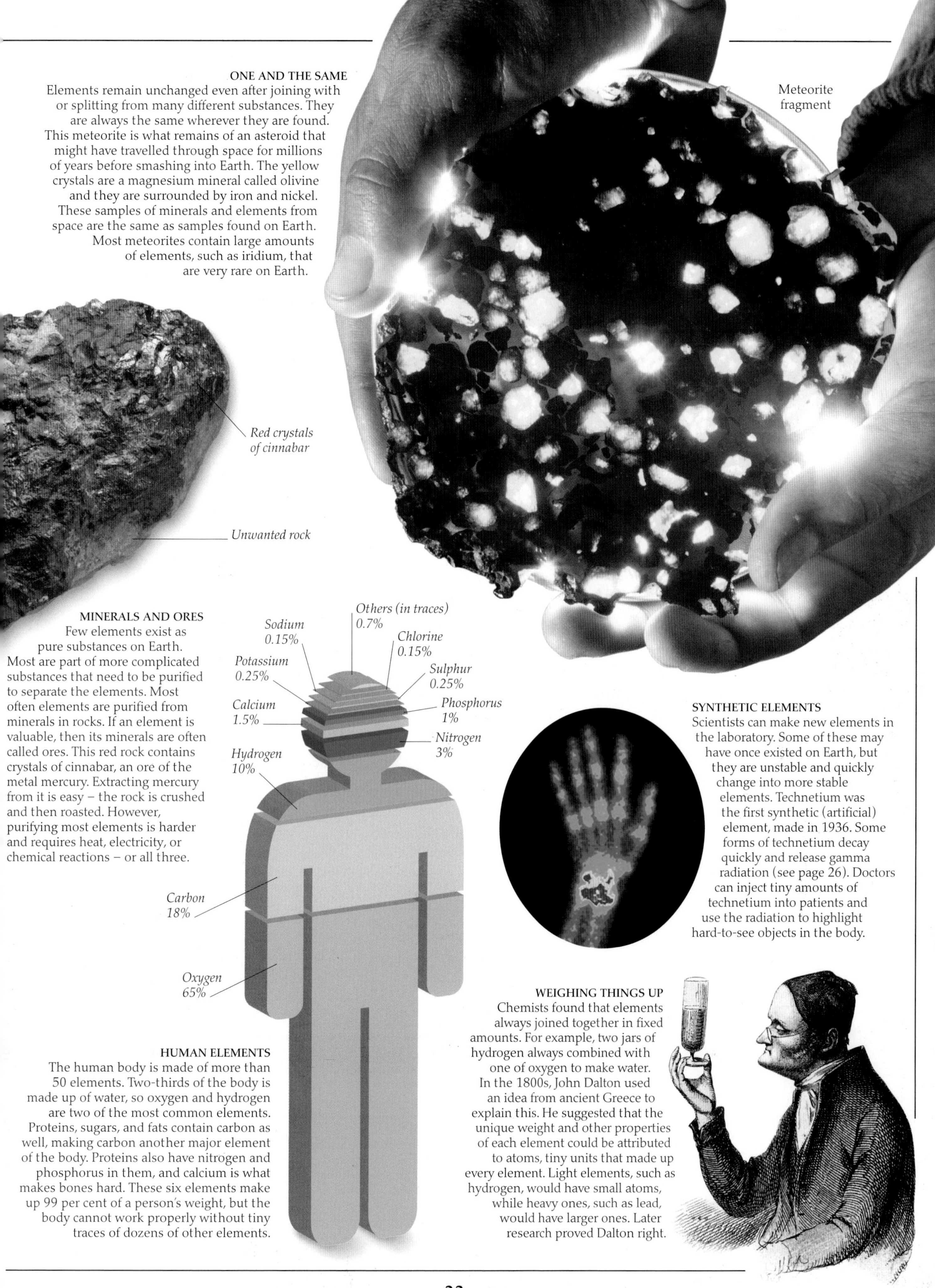

ONE AND THE SAME

Elements remain unchanged even after joining with or splitting from many different substances. They are always the same wherever they are found. This meteorite is what remains of an asteroid that might have travelled through space for millions of years before smashing into Earth. The yellow crystals are a magnesium mineral called olivine and they are surrounded by iron and nickel. These samples of minerals and elements from space are the same as samples found on Earth. Most meteorites contain large amounts of elements, such as iridium, that are very rare on Earth.

MINERALS AND ORES

Few elements exist as pure substances on Earth. Most are part of more complicated substances that need to be purified to separate the elements. Most often elements are purified from minerals in rocks. If an element is valuable, then its minerals are often called ores. This red rock contains crystals of cinnabar, an ore of the metal mercury. Extracting mercury from it is easy – the rock is crushed and then roasted. However, purifying most elements is harder and requires heat, electricity, or chemical reactions – or all three.

SYNTHETIC ELEMENTS

Scientists can make new elements in the laboratory. Some of these may have once existed on Earth, but they are unstable and quickly change into more stable elements. Technetium was the first synthetic (artificial) element, made in 1936. Some forms of technetium decay quickly and release gamma radiation (see page 26). Doctors can inject tiny amounts of technetium into patients and use the radiation to highlight hard-to-see objects in the body.

HUMAN ELEMENTS

The human body is made of more than 50 elements. Two-thirds of the body is made up of water, so oxygen and hydrogen are two of the most common elements. Proteins, sugars, and fats contain carbon as well, making carbon another major element of the body. Proteins also have nitrogen and phosphorus in them, and calcium is what makes bones hard. These six elements make up 99 per cent of a person's weight, but the body cannot work properly without tiny traces of dozens of other elements.

WEIGHING THINGS UP

Chemists found that elements always joined together in fixed amounts. For example, two jars of hydrogen always combined with one of oxygen to make water. In the 1800s, John Dalton used an idea from ancient Greece to explain this. He suggested that the unique weight and other properties of each element could be attributed to atoms, tiny units that made up every element. Light elements, such as hydrogen, would have small atoms, while heavy ones, such as lead, would have larger ones. Later research proved Dalton right.

Inside atoms

Circular magnet creates magnetic field

Rays are emitted inside glass tube

Cathode rays are bent

ELEMENTS ARE MADE UP of indivisible particles called atoms. This idea was first suggested about 2,400 years ago by the Greek philosopher Democritus. He also put forward the idea that rough atoms gripped each other while smoother ones did not. Scientists found that atoms of some elements are heavier than those of others. Atoms of different elements also join together to make other substances, such as water. In the 20th century, scientists determined that atoms are composed of tinier particles – protons, neutrons, and electrons. This discovery revealed how atoms really work.

BENDING RAYS

In 1897, J J Thomson experimented with cathode rays – beams created when electricity passes through an almost airless tube. Thomson found he could move the rays with electricity and magnets. He concluded that these rays contained particles that were so small and light that they must have come from inside atoms. Scientists later named Thomson's particles electrons.

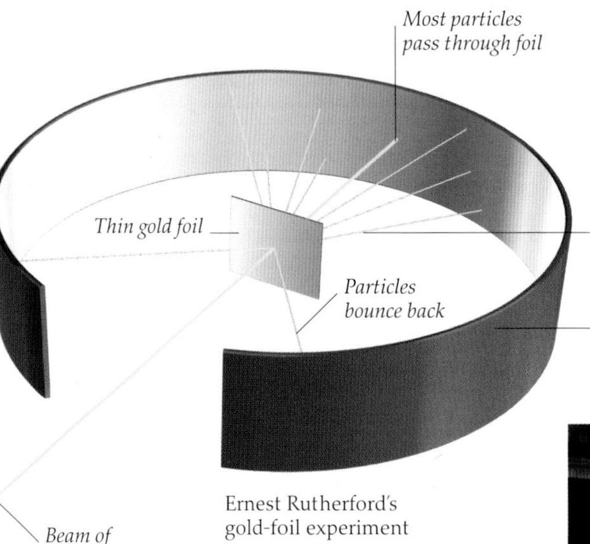

Ernest Rutherford's gold-foil experiment

INSIDE AN ATOM

Electrons are negatively charged, and scientists knew that their charges (see page 40) must be balanced by positive charges inside an atom. Atoms are neutral (neither positive nor negative). Opposite charges attract each other, while similar charges repel one another. Scientists knew about positively charged alpha particles, but they did not know if they came from atoms. In 1909, Ernest Rutherford discovered the positive parts of an atom, by firing alpha particles at gold foil. Nearly all of them went straight through, but a tiny number bounced back. Rutherford realized that they were being repelled by something – a tiny positively charged nucleus at the very centre of atoms. He reasoned that the nucleus was a ball of charged particles he named protons.

Molten iron being rolled in a mill

EMPTY SPACE

Even though atoms are tiny – 10 million hydrogen atoms lined up would measure less than 1 mm (0.04 in) – they contain mostly empty space. The nucleus is 10,000 times smaller than the whole atom. If an atom was the size of the Colosseum – the giant arena in Rome, Italy – its nucleus would be the size of a pea. The electrons are negatively charged and move throughout the atom, held in orbit by the positive nucleus. Electrons are 1,600 times less massive (heavy) than protons, so the atom is mainly empty space. Electrons move so fast, though, that they seem to fill up this space.

The Colosseum

TOUGH METALS

Electrons are arranged in sphere shapes or layers called shells around an atom's nucleus. In metal atoms, the outermost shell is almost empty, with only an electron or two. These escape atoms easily and form a sea of electrons that are shared by all the atoms of the metal, glueing them together strongly. This "electron glue" makes a metal tough, so it can be rolled flat or moulded into shapes without breaking.

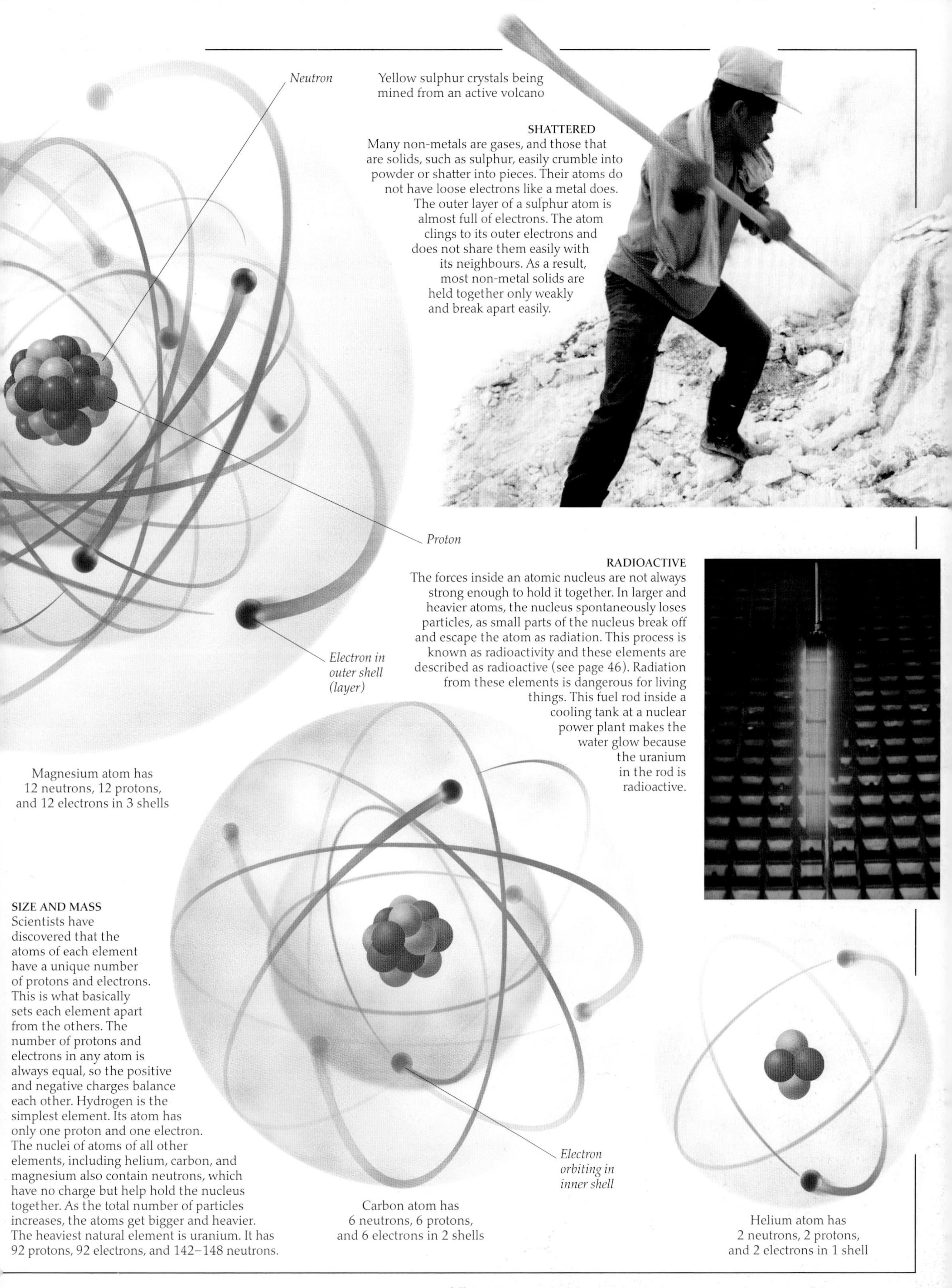

Yellow sulphur crystals being mined from an active volcano

SHATTERED

Many non-metals are gases, and those that are solids, such as sulphur, easily crumble into powder or shatter into pieces. Their atoms do not have loose electrons like a metal does. The outer layer of a sulphur atom is almost full of electrons. The atom clings to its outer electrons and does not share them easily with its neighbours. As a result, most non-metal solids are held together only weakly and break apart easily.

RADIOACTIVE

The forces inside an atomic nucleus are not always strong enough to hold it together. In larger and heavier atoms, the nucleus spontaneously loses particles, as small parts of the nucleus break off and escape the atom as radiation. This process is known as radioactivity and these elements are described as radioactive (see page 46). Radiation from these elements is dangerous for living things. This fuel rod inside a cooling tank at a nuclear power plant makes the water glow because the uranium in the rod is radioactive.

Magnesium atom has 12 neutrons, 12 protons, and 12 electrons in 3 shells

SIZE AND MASS

Scientists have discovered that the atoms of each element have a unique number of protons and electrons. This is what basically sets each element apart from the others. The number of protons and electrons in any atom is always equal, so the positive and negative charges balance each other. Hydrogen is the simplest element. Its atom has only one proton and one electron. The nuclei of atoms of all other elements, including helium, carbon, and magnesium also contain neutrons, which have no charge but help hold the nucleus together. As the total number of particles increases, the atoms get bigger and heavier. The heaviest natural element is uranium. It has 92 protons, 92 electrons, and 142–148 neutrons.

Carbon atom has 6 neutrons, 6 protons, and 6 electrons in 2 shells

Helium atom has 2 neutrons, 2 protons, and 2 electrons in 1 shell

Radio waves | Microwaves | Heat (infrared) | Ultraviolet | X-rays | Gamma rays

Visible light

Electromagnetic spectrum – the sequence of all possible forms of electromagnetic radiation

WAVES
Electromagnetic waves move at the speed of light, but they carry varying amounts of energy, depending on the length of their waves – their "wavelength". Wavelength is the distance from the top of one wave to the top of the next. Low-energy radio waves have waves 10 m (33 ft) long or more. At the other end of the spectrum, gamma rays contain so much energy that billions of their waves fit into a millimetre!

Making waves

ONE OF THE GREAT DISCOVERIES OF THE 20TH CENTURY was how atoms produce light, heat, and other invisible rays. The electrons around an atom are not fixed in place and each electron can collect or give out energy. An electron jumps away from the nucleus when it takes in energy, and when it settles back into its former position, it emits energy. This energy travels as waves called electromagnetic radiation. The most familiar forms of this radiation are light, which we see, and heat, which we feel. However, there are other invisible forms of electromagnetic radiation, including ultraviolet rays, X-rays, and radio waves.

RELEASING ENERGY
The heat and light from this fire are released by atoms inside the burning wood. Burning is a chemical reaction in which oxygen combines with the chemicals in wood (see page 37). As the atoms in the wood form new compounds with the oxygen atoms, they tend to lose energy and become more stable. The lost energy is released as radiation.

COLD LIGHT
Bulbs and flames produce light with heat. However, it is not necessary for heat to be produced along with light. Some animals produce pure light inside their bodies, without generating heat at the same time. Deep-sea creatures, such as this jellyfish, live in dark water and use the light emitted by their bodies to attract mates or lure prey. The light is made by oxygen reacting with certain light-emitting chemicals inside special organs. The reaction releases radiation as "cold" light. If the reaction released a lot of heat energy, it would cook the animal alive!

White light – a mixture of different wavelengths

Glass prism splits the beam of white light

White light split into the component wavelengths, forming a spectrum of colours

SPLITTING LIGHT
When a beam of white light shines into a prism, it comes out of the other side as a rainbow of colours. White light is made up of waves of different lengths. Our eyes and brains can distinguish between the lengths of these waves, which gives us the sensation of different colours. Red light has the longest wavelength, while violet light has the shortest. Isaac Newton described the colours as a "spectrum" – the word now used for the whole range of electromagnetic radiation. Newton's spectrum refers to the visible part of the electromagnetic spectrum and consists of the colours from red to violet. When waves of different lengths are mixed, they form white light.

GOING WIRELESS
In 1901, Italian inventor Guglielmo Marconi showed that signals carried by radio waves could be made to travel thousands of miles, by bouncing the waves off a layer of charged particles high in the atmosphere. Here, Marconi and his team are seen setting up a radio aerial to receive the radio signals. Radio waves may carry tiny amounts of energy, but they are useful. A radio signal carries information by varying its amplitude – the height of the wave – or its wavelength. Today, radio waves are used to carry signals to mobile phones, televisions, or even space probes.

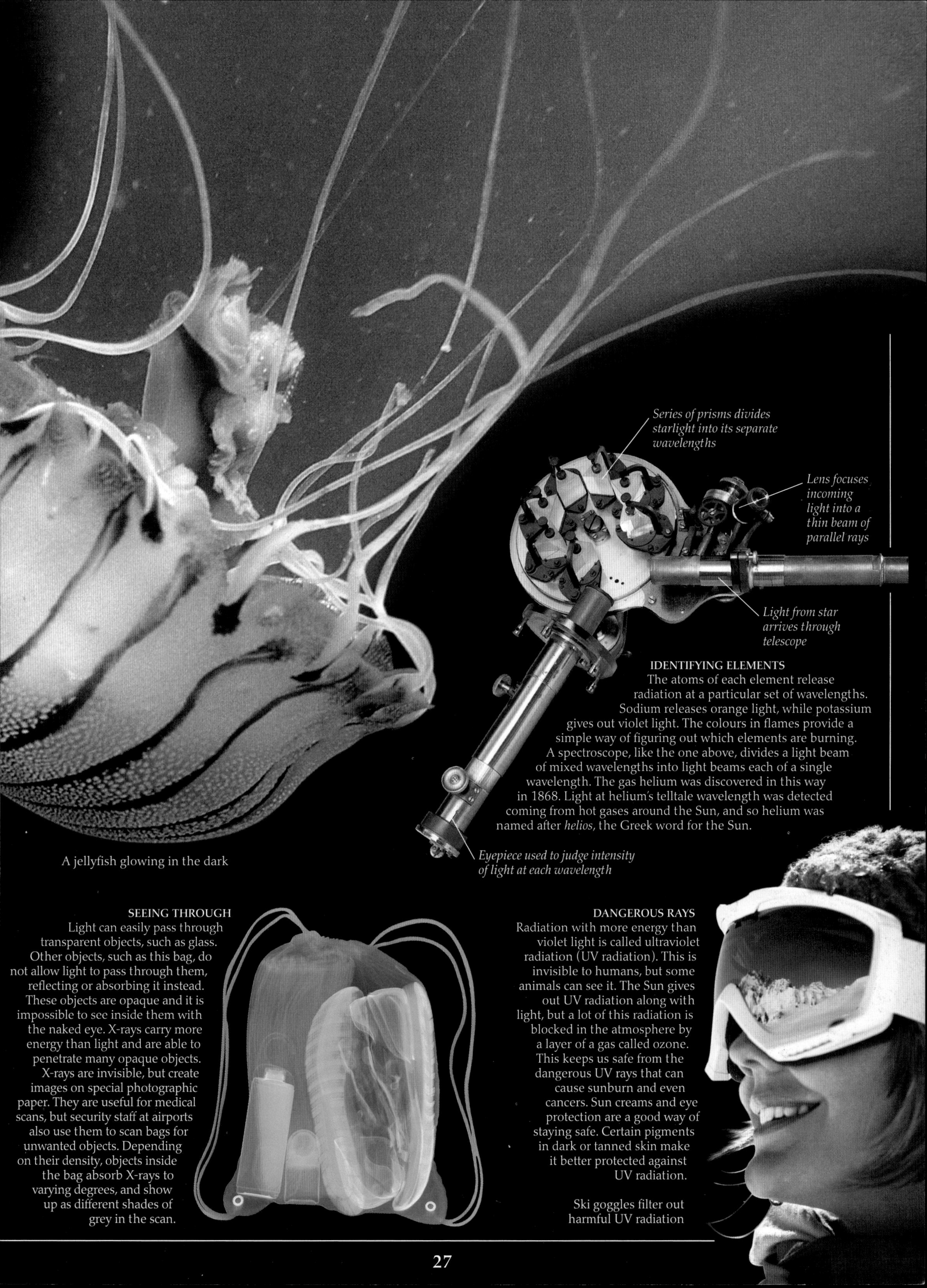

A jellyfish glowing in the dark

IDENTIFYING ELEMENTS
The atoms of each element release radiation at a particular set of wavelengths. Sodium releases orange light, while potassium gives out violet light. The colours in flames provide a simple way of figuring out which elements are burning. A spectroscope, like the one above, divides a light beam of mixed wavelengths into light beams each of a single wavelength. The gas helium was discovered in this way in 1868. Light at helium's telltale wavelength was detected coming from hot gases around the Sun, and so helium was named after *helios*, the Greek word for the Sun.

SEEING THROUGH
Light can easily pass through transparent objects, such as glass. Other objects, such as this bag, do not allow light to pass through them, reflecting or absorbing it instead. These objects are opaque and it is impossible to see inside them with the naked eye. X-rays carry more energy than light and are able to penetrate many opaque objects. X-rays are invisible, but create images on special photographic paper. They are useful for medical scans, but security staff at airports also use them to scan bags for unwanted objects. Depending on their density, objects inside the bag absorb X-rays to varying degrees, and show up as different shades of grey in the scan.

DANGEROUS RAYS
Radiation with more energy than violet light is called ultraviolet radiation (UV radiation). This is invisible to humans, but some animals can see it. The Sun gives out UV radiation along with light, but a lot of this radiation is blocked in the atmosphere by a layer of a gas called ozone. This keeps us safe from the dangerous UV rays that can cause sunburn and even cancers. Sun creams and eye protection are a good way of staying safe. Certain pigments in dark or tanned skin make it better protected against UV radiation.

Ski goggles filter out harmful UV radiation

Rays of light

THE SCIENCE THAT EXPLAINS how light behaves is called optics. The difference between light and other types of electromagnetic radiation is that its wavelengths can be seen by people and animals. Chemicals in the eyes are sensitive to different wavelengths in light bouncing off objects all around us. Eyes function as sensors, collecting this information for the brain to convert into pictures. However, optics is not just about what we see. The study of how light interacts with different substances is useful in technology. High-speed communication cables send information as flickering beams of light bouncing down transparent glass wires known as optical fibres, while a DVD player reads a video disk by the way it reflects a laser.

SHIFTING IMAGE
Sometimes light appears to play tricks. Light changes direction slightly as it moves between different materials, such as air and water. This property of light is called refraction. Thanks to this shift, the light bouncing off these feet appears to come from the right of where they really are – so the feet appear disconnected from the legs.

BOUNCING BACK
One of the most familiar things light does is reflect. A beam of light bounces off the surface of a substance to create a reflection. The reflected light creates an image of the substance in our eyes. This is a real image. If a substance has a smooth surface, such as a mirror, every beam of light bounces off in the same direction, creating an image of where the light came from originally. The mirror image looks as though it comes from behind the mirror. It is always back to front and appears the same distance away as the reflected object. This image is virtual. The sculpture seen here is a real image, while the Chicago skyline seen on it is virtual.

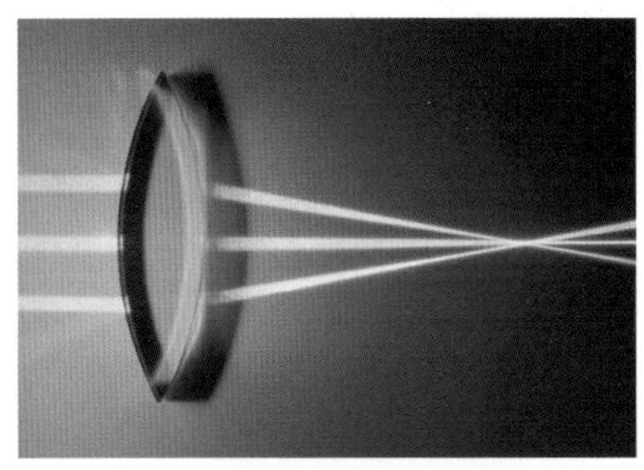

Convex lens focuses beams of light

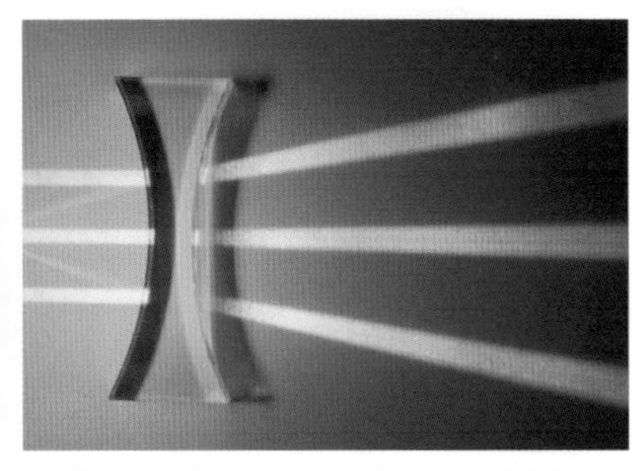

Concave lens spreads beams of light

CURVING LIGHT

A lens is built to refract the light that shines through it. However, its curved surface makes the light refract at different angles. Convex lenses bend light beams so they are focused to a single point. Concave lenses do the opposite, spreading the beams apart. Objects viewed through a concave lens appear to be closer, or bigger, than they really are, because the light passing through them has been made to fill a larger area.

Black ball absorbs all the light

Red ball absorbs all colours of light except red

White ball reflects all the light and absorbs nothing

COLOUR QUESTION

A coloured object appears so because of the way it reflects the light that shines on it. Sunlight is a mixture of all colours, which add up to form brilliant white light. A white ball looks that way because it reflects every shade of light that hits it. A black ball does not reflect any light at all, absorbing all of it instead. A coloured object, such as a red ball, reflects only that colour, which has a particular wavelength (see pages 26–27). Every other colour is absorbed.

LASER BEAM

Natural light is a tangle of waves of different wavelengths, all shining in different directions. Laser light is quite the opposite. A laser beam normally contains light of a single colour, with all the waves lined up perfectly. The waves are regular and in step with each other, and behave in identical ways. This is why lasers form straight beams, which reflect and refract precisely and make this laser light show look spectacular. But lasers are used for a lot more than entertainment. They are used in bar-code readers, in surgeries, and as super-hot cutting tools. A laser beam is even bounced off the Moon to measure exactly how far it is from Earth.

Chicago's skyscrapers reflected on the curved mirror surface of a public sculpture

BLUE SKY

Air appears to be colourless until you look up, when the sky seems blue. On its way to the ground, some of the light from the Sun hits the gas particles in the atmosphere and is scattered in all directions. The colour blue gets scattered more than the other colours in sunlight. When you look up, most of what can be seen is the blue light scattered across the sky.

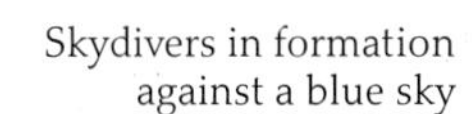

Skydivers in formation against a blue sky

Sound system

SOUND WAVES ARE INVISIBLE VIBRATIONS that travel through air, water, and other substances. Our ears pick up these vibrations. Just like light waves, sounds can be reflected and refracted. Sound waves also have a wavelength. Low hums have long wavelengths, while short-wavelength sounds are high-pitched squeaks. Sounds are also described by their frequency. This is a measure of how many wavelengths the sound wave moves through in one second. Sounds with long wavelengths have a low frequency and those with short wavelengths have a high frequency. Unlike light and heat, sound needs a medium (something to carry it), such as air, water, or even rock. Space is silent as there is no medium to carry the sound waves.

Jet fighter breaking the sound barrier

Thicker strings make deeper notes

Hole through which sound is released

Bow made from horse hairs

MUSIC TO YOUR EARS
Sound is formed when air is disturbed, sending a wave out in all directions. Most sounds are a jumble of many frequencies, so they sound like rustles and thuds. Musical instruments produce sound at a single frequency, creating a note of music. Drawing a bow across the strings of a violin creates a vibration, which passes to the air inside the wooden case. This begins to vibrate at the same frequency. The sound waves produced emerge through the curved holes on the violin.

Jar is more than 1 m (3 ft) wide

Dragon jaw holds ball

Ball falls into mouth of toad showing direction of earthquake

DETECTING QUAKES
Vibrations inside Earth are called seismic waves. These large-scale movements of rocks in Earth's crust send out powerful waves, which cause earthquakes once they reach the surface. Scientists record seismic waves to work out what is happening underground. In 132 CE, Chinese inventor Zhang Heng built this detector to pick up the faint seismic waves preceding earthquakes. A pendulum inside the jar is linked to balls in the jaws of the dragons. Seismic waves swing the pendulum, which makes a ball fall into the mouth of a toad, indicating the direction from which an earthquake is approaching.

Compression Rarefaction

IN AND OUT WAVES
Sound waves do not rise and fall like the waves of water approaching a beach. Instead, a sound wave passing through air is made up of parts where the air molecules are squeezed together (compression) and parts where the air molecules are stretched apart (rarefaction). A wave is moving along this spring in a similar way. The air does not actually travel along with the wave. Small parcels of air just jump backwards and forwards a little as the wave passes.

SONIC BOOM
The speed of sound in air is about 1,225 kph (760 mph). When a jet flies, it pushes the air out of the way. This creates a shock wave, which grows bigger as the jet accelerates to the speed of sound. The shock wave is just air and cannot move faster than sound. The jet breaks the sound barrier when it flies faster than the speed of sound, powering through the shock wave and leaving it behind. The shock wave then collapses, creating a wave of sound, a rumbling "sonic boom" that can be heard far away.

KEEPING QUIET
When a sound reflects off a surface, it is heard as an echo. There are echoes all around, but most of them are too faint to be heard. When scientists study sounds, they need to get rid of echoes. They use an anechoic chamber, with walls covered in soft shapes that absorb all the echoes. In 2005, an anechoic chamber in Minneapolis, US, was awarded the world record for being the quietest place on Earth.

Large, sensitive ears amplify sounds received

SEEING WITH SOUND
Bats use sound to find their way around when it is too dark to see. A bat sends out high-pitched chirps that echo off the objects around it. By listening to the echoes, the bat can tell exactly where an object is, and may even make out details of the object's shape. This is called echolocation. A bat's calls are too high in frequency for humans to hear, but they are very powerful. Using specialized muscles, the animal disengages its sensitive ears when it calls, otherwise the loud sounds would deafen it.

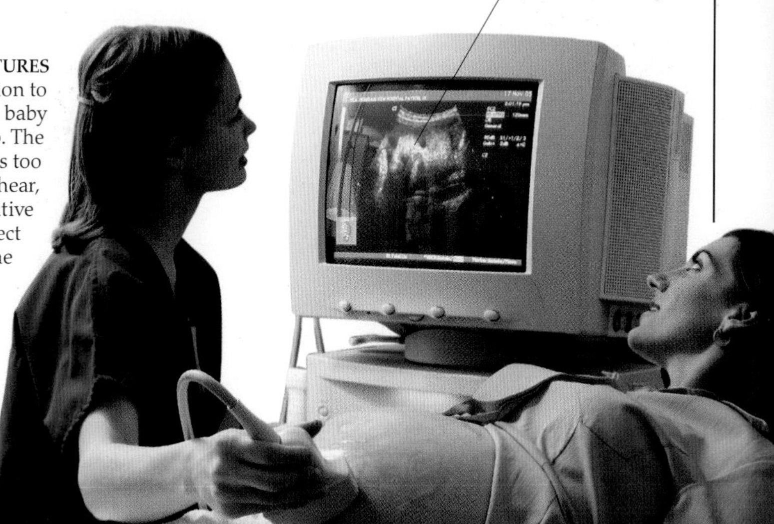

Echoes from body parts create images

SOUND PICTURES
Ultrasound machines use echolocation to look inside a person's body or at a baby growing inside its mother's womb. The sound produced by the machine is too high in frequency for people to hear, but a probe in the machine is sensitive enough to detect the echoes that reflect off organs or other objects inside the body. A computer uses these echoes to create a picture of the body's interior on a screen. This system is safe because, unlike X-rays, the sound waves do not contain enough energy to damage tissues they travel through.

States of matter

ATOMS ARE NEVER STILL. Even the atoms in a solid object – which always has a fixed shape – are vibrating back and forth. As the object gets hotter, the atoms vibrate faster. The stronger vibrations pull on the bonds holding the atoms together. Eventually the bonds will begin to break. Without enough bonds to keep it solid, a substance such as ice melts into a liquid. Most of the atoms are still connected, but there are not enough bonds to give the liquid a fixed shape. As the heating continues, all the bonds break and the atoms (or groups of atoms) float away on their own. The liquid has boiled into a gas. This process also runs in reverse. As atoms cool, they link together again. A gas condenses into droplets of liquid, which then freeze solid. Solid, liquid, and gas are the main states of matter.

Water vapour is an invisible gas

Solid ice

Liquid water

TAKING THE TEMPERATURE
Anything can melt, boil, or freeze. When heat is applied to this ice sculpture, the solid ice melts into droplets of liquid water. Heating the water makes it boil into gaseous water vapour. This then condenses in the cold air to form steam, which is floating liquid droplets. Scientists use water as a standard to measure temperature – how much heat is in something. Water freezes and boils at fixed temperatures. Scientists called its freezing point 0 degrees on the Celsius scale. Water's boiling point is at 100°C (212°F). The temperature of everything else is compared to these points. For example, the temperature of a human body is about 37°C (98.6°F).

SPREADING OUT
Gas particles are not connected to anything, and move rapidly in all directions, until they encounter and bounce off a surface. This process of spreading out is called diffusion and it is why the smoke in this parachute display is filling the sky. Diffusion also explains how smells spread across a room, and how gases move from a breath of air inside the lungs into the body's blood stream.

Foggy mixture of carbon dioxide and condensed water vapour

Water warms up dry ice, speeding its sublimation

LIQUID FREE
A few solids turn directly into gases, without becoming liquids. This process is called sublimation. The most familiar example of this is dry ice, which is used to create heavy fog in theatres and magic shows. Dry ice is frozen carbon dioxide. It is much colder than ice made from water and has no liquid state, turning instead into carbon dioxide gas at room temperature. Putting it in water, as shown here, speeds up this process. The cold carbon dioxide coming out of the flask comes into contact with water vapour in the air, making it condense into a thick, white fog. This sinks to the ground because carbon dioxide gas is heavier than air.

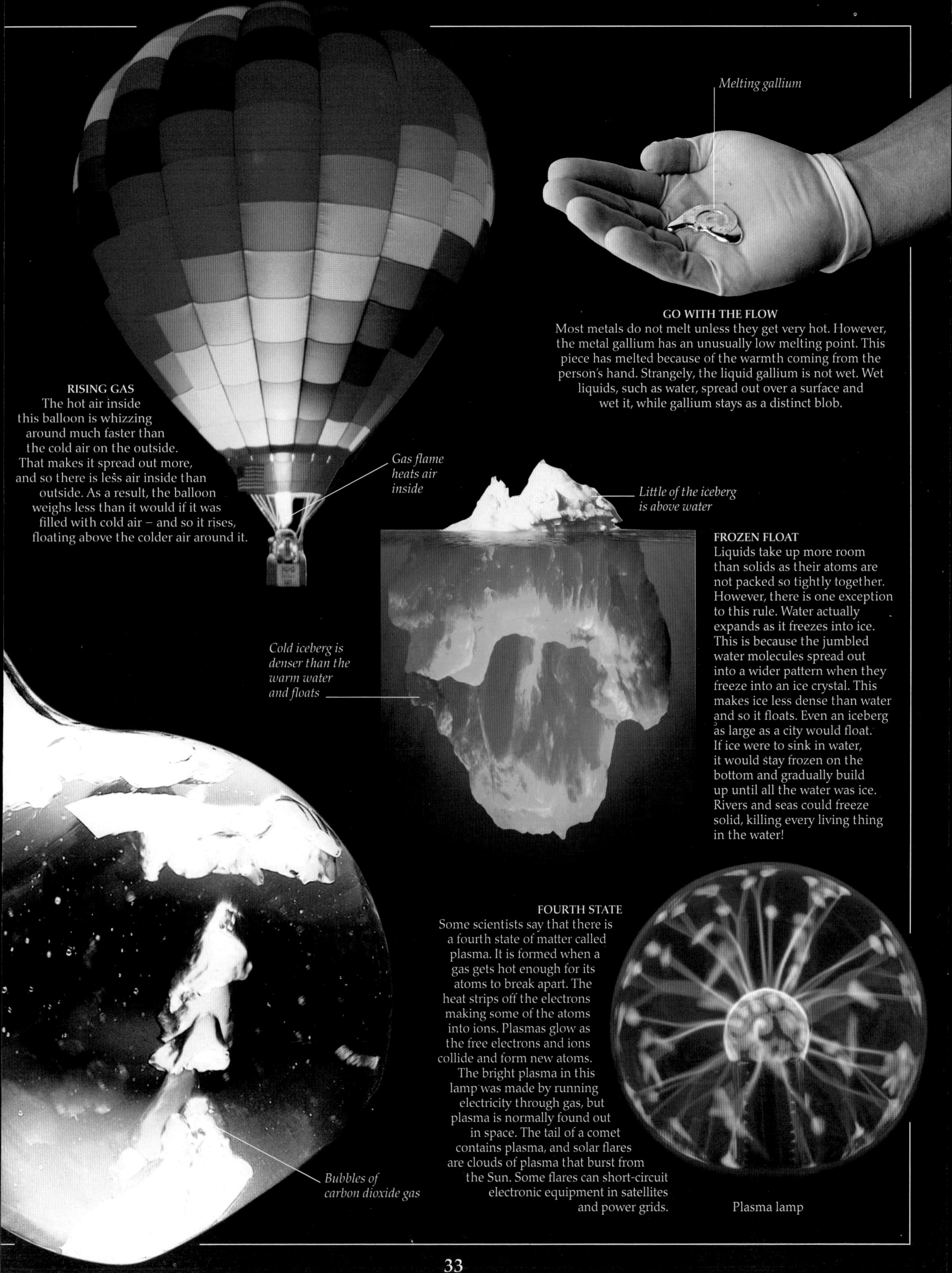

RISING GAS

The hot air inside this balloon is whizzing around much faster than the cold air on the outside. That makes it spread out more, and so there is less air inside than outside. As a result, the balloon weighs less than it would if it was filled with cold air – and so it rises, floating above the colder air around it.

Gas flame heats air inside

Melting gallium

GO WITH THE FLOW

Most metals do not melt unless they get very hot. However, the metal gallium has an unusually low melting point. This piece has melted because of the warmth coming from the person's hand. Strangely, the liquid gallium is not wet. Wet liquids, such as water, spread out over a surface and wet it, while gallium stays as a distinct blob.

Little of the iceberg is above water

Cold iceberg is denser than the warm water and floats

FROZEN FLOAT

Liquids take up more room than solids as their atoms are not packed so tightly together. However, there is one exception to this rule. Water actually expands as it freezes into ice. This is because the jumbled water molecules spread out into a wider pattern when they freeze into an ice crystal. This makes ice less dense than water and so it floats. Even an iceberg as large as a city would float. If ice were to sink in water, it would stay frozen on the bottom and gradually build up until all the water was ice. Rivers and seas could freeze solid, killing every living thing in the water!

FOURTH STATE

Some scientists say that there is a fourth state of matter called plasma. It is formed when a gas gets hot enough for its atoms to break apart. The heat strips off the electrons making some of the atoms into ions. Plasmas glow as the free electrons and ions collide and form new atoms. The bright plasma in this lamp was made by running electricity through gas, but plasma is normally found out in space. The tail of a comet contains plasma, and solar flares are clouds of plasma that burst from the Sun. Some flares can short-circuit electronic equipment in satellites and power grids.

Bubbles of carbon dioxide gas

Plasma lamp

Mixing it up

TYPES OF MIXTURE
These marbles are mixtures of coloured glass that were swirled together as hot liquids before being cooled into solids. There are many kinds of mixture. A solid or gas dissolved in a liquid forms a mixture called a solution. A liquid mixed into a solid forms a wobbly gel – like fruit jelly. A gas mixed into a solid creates foam, while a liquid mixed into a gas makes an aerosol spray.

MANY COMMON SUBSTANCES are actually mixtures of several separate ingredients. Mixtures are either even or uneven in composition. An example of an even mixture is sea water, made from salt and water. Salty water looks the same as pure water, because the salt dissolves – spreading out evenly in the water and becoming invisible. The same is not true of mud, an uneven mixture. In mud, blobs of soil just float in water – they are not evenly spread through it. So the soil is still visible, and the mixture looks like a combination of its ingredients.

MIXTURE OF MIXTURES
Ice cream is difficult to make at home because it is a complex mixture of sugar, flavourings, water, cream, and air. The water is actually a half-frozen mix of water and ice, with sugar and flavourings dissolved in it to produce the sweet taste. Cream does not dissolve, but breaks up into tiny blobs of fat that blend with the water creating a cloudy mixture called an emulsion. Bubbles of air are stirred in to make this gooey emulsion softer and more pleasant to eat.

A scoop of mango ice cream

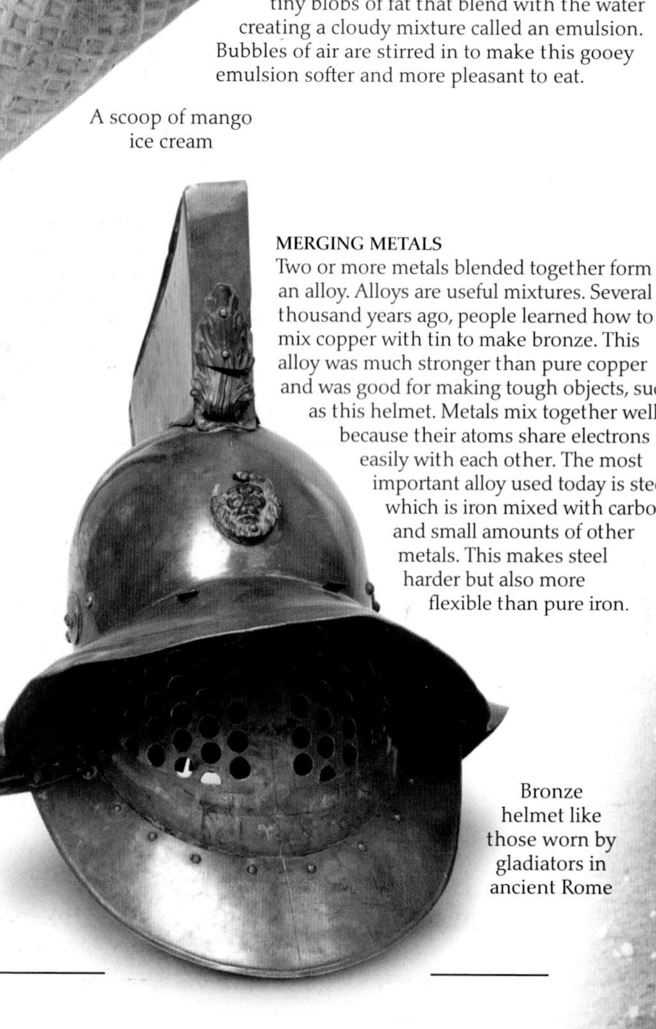

MERGING METALS
Two or more metals blended together form an alloy. Alloys are useful mixtures. Several thousand years ago, people learned how to mix copper with tin to make bronze. This alloy was much stronger than pure copper and was good for making tough objects, such as this helmet. Metals mix together well because their atoms share electrons easily with each other. The most important alloy used today is steel, which is iron mixed with carbon and small amounts of other metals. This makes steel harder but also more flexible than pure iron.

Bronze helmet like those worn by gladiators in ancient Rome

PURIFYING PROCESS

The ingredients in a mixture are not joined together – just spread out among each other. A mixture can thus be split into its ingredients quite easily. For example, a mixture such as a solution can be purified by heating it till the liquid portion boils away leaving the solid ingredients behind. This is how salt is commonly obtained from sea water. These workers in Kerala, India, are collecting crystals of pure sea salt, left in shallow ponds after the sea water has evaporated in the Sun.

ONE WAY THROUGH

Filtration separates ingredients in an uneven mixture based on the sizes of the different ingredients. Filters can be made of paper, cloth, or metal – anything that has a lot of tiny holes. One ingredient can pass through the small holes in the filter, but the other ingredients cannot. Filters are used in many ways. Muslin cloth is used to separate solids from solutions in experiments, while beds of fine sand are used to filter out the solid waste from sewage water.

WON'T MIX, CAN'T MIX

Some things never mix. Oil in water floats up to the surface and forms a separate layer on top. The oil and water do not blend, so they stay pure even when both are put together. This lava lamp is filled with water and wax – as a pair, these are immiscible (non-mixing) substances. The lamp heats the wax, which melts and moves around in the water, never mixing with it.

ON THE MOVE

Different ingredients in a mixture move through a substance at different speeds. This principle is used in chromatography for separating mixtures containing similar ingredients that are hard to separate. Dark ink, for instance, contains several colours of dye. The dyes soak through blotting paper at their own speeds, so each dye travels a certain distance, creating a band of single colour.

Chemical reactions

ATOMS OF MOST ELEMENTS COMBINE with atoms of other elements to form new substances known as compounds. These have different properties from the elements that combined to make them. For example, atoms of hydrogen and oxygen – two gases – react to make liquid water. Compounds form during chemical reactions, when atoms rearrange themselves into different combinations. The atoms in a compound are chemically bonded together. They form bonds because atoms are more stable when connected. Some chemical bonds involve atoms sharing electrons. Other bonds form when one atom gives electrons to another.

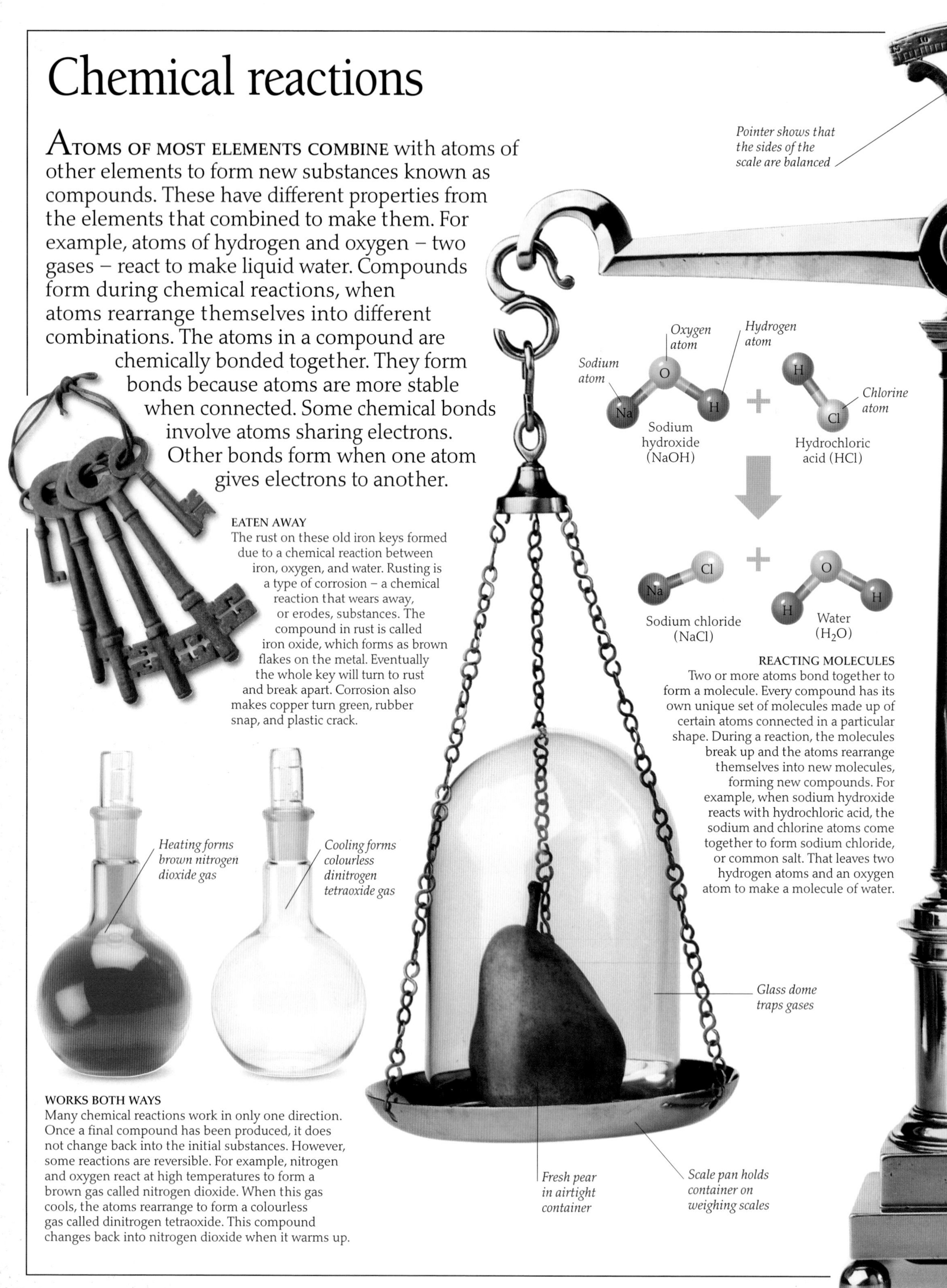

EATEN AWAY
The rust on these old iron keys formed due to a chemical reaction between iron, oxygen, and water. Rusting is a type of corrosion – a chemical reaction that wears away, or erodes, substances. The compound in rust is called iron oxide, which forms as brown flakes on the metal. Eventually the whole key will turn to rust and break apart. Corrosion also makes copper turn green, rubber snap, and plastic crack.

REACTING MOLECULES
Two or more atoms bond together to form a molecule. Every compound has its own unique set of molecules made up of certain atoms connected in a particular shape. During a reaction, the molecules break up and the atoms rearrange themselves into new molecules, forming new compounds. For example, when sodium hydroxide reacts with hydrochloric acid, the sodium and chlorine atoms come together to form sodium chloride, or common salt. That leaves two hydrogen atoms and an oxygen atom to make a molecule of water.

WORKS BOTH WAYS
Many chemical reactions work in only one direction. Once a final compound has been produced, it does not change back into the initial substances. However, some reactions are reversible. For example, nitrogen and oxygen react at high temperatures to form a brown gas called nitrogen dioxide. When this gas cools, the atoms rearrange to form a colourless gas called dinitrogen tetraoxide. This compound changes back into nitrogen dioxide when it warms up.

NOTHING CREATED OR DESTROYED

Atoms are not created or destroyed during chemical reactions. The number of atoms at the end of a reaction is always the same as it was before the reaction. The atoms are just bonded to new neighbours. In this imaginary experiment, a pear weighs the same when fresh as when it has rotted away into a mouldy sludge due to many different chemical reactions. Much of the rotten pear has turned into gases, but these still contribute to the overall weight.

HEAT NEEDED

Many chemical reactions can only start when the ingredients are hot. Iron and sulphur need to be heated together to make them react to form iron sulphide. Once enough heat energy has been provided, the reaction will start and begin to give out energy. The iron sulphide reaction actually gives out more energy than it initially uses. Scientists call this an exothermic reaction. The products end up being very hot. Other reactions are endothermic – they release less energy than that used to make them start. These reactions may even take heat from the surroundings, causing a drop in temperature.

A HELPING HAND

Some reactions occur in a fraction of a second, but others may continue for years. Heat normally makes reactions run a bit faster. However, some reactions are so slow that they would hardly happen without help provided by substances called catalysts. A catalyst is not used up in a reaction, but only brings the ingredients together so they can react. Bread dough rises thanks to yeast – a fungus that acts as a catalyst. Yeast speeds up the conversion of the sugars in flour into water and carbon dioxide gas, which makes the dough puff up.

BURNING ISSUE

Oxygen is one of the most reactive elements. It often reacts with compounds and other elements with a burning reaction called combustion, giving out light and heat, which appear as flames. An industrial oxyacetylene torch uses oxygen and burns at 3,500°C (6,300°F). It melts metal pieces so they can be welded into a single unit. Combustion is perhaps the most useful chemical reaction. Humans have used fire for more than 500,000 years to keep warm and cook food. Today, we still rely on combustion to release energy from the fuels used to drive cars and generate electricity in power stations.

Acids and alkalis

CHEMICAL REACTIONS HAPPEN all around us, not only in laboratories. Many of them involve chemicals called acids. Acid molecules exist in different shapes and sizes, but all have a hydrogen atom that breaks off easily to attack other chemicals. Acids are often dissolved in water, and when the hydrogen atom breaks free, it becomes a highly reactive ion – an atom missing its electron. Acids react most strongly with alkalis, which are chemicals that release hydroxide ions, made from an oxygen and a hydrogen atom. The hydrogen and hydroxide ions react to form water. The rest of the acid bonds with the alkali, producing a chemical called a salt.

SOUR NOT SWEET
Acids are common in nature. Lemon juice tastes sour because it contains citric acid. This weak acid releases few hydrogen ions and is safe to swallow. Some natural acids are strong enough to react with skin, making it sore. A nettle sting is caused by tiny amounts of strong acid on the hairy leaves.

CLEANING CHEMICALS
Soap is made by an acid–alkali reaction. Soap makers combine fatty acids (complex acids in plant and animal oils and fats) with a strong alkali, such as sodium hydroxide. These react to form soap – an alkaline mixture of salts. The salt molecules have two parts. The part that dissolves in water is hydrophilic, meaning it "loves water". The other part is hydrophobic, meaning it "hates water", and only mixes with grease and dirt. Soap cleans by collecting lumps of dirt with the hydrophobic end. Then the hydrophilic part mixes into water and the whole thing is washed away. Synthetic detergents, such as washing powder, work in a similar way.

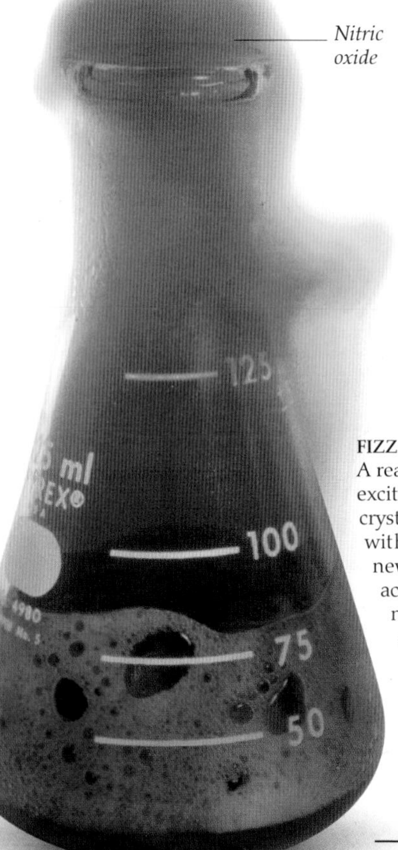

FIZZ BANG
A reaction involving an acid is often exciting, with fizzing gases and colourful crystals. The reaction normally ends with the formation of two or three new chemicals. In this flask, nitric acid is reacting with copper. Orange nitric oxide gas bubbles out, and the copper disappears as it forms a green salt called copper nitrate. It is very different from the salt in food, but both are formed by an acid reaction.

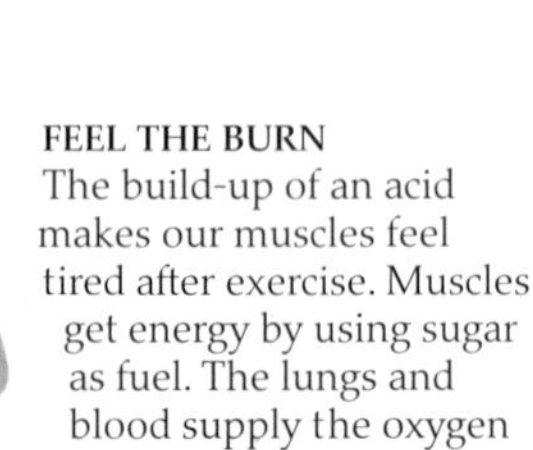

FEEL THE BURN
The build-up of an acid makes our muscles feel tired after exercise. Muscles get energy by using sugar as fuel. The lungs and blood supply the oxygen needed to release energy from sugar. When the muscles have to work hard, the lungs cannot provide enough oxygen to them, and sugar is processed without oxygen. This oxygen-free process turns the sugar into lactic acid. When the acid builds up, it creates a burning sensation inside tired muscles.

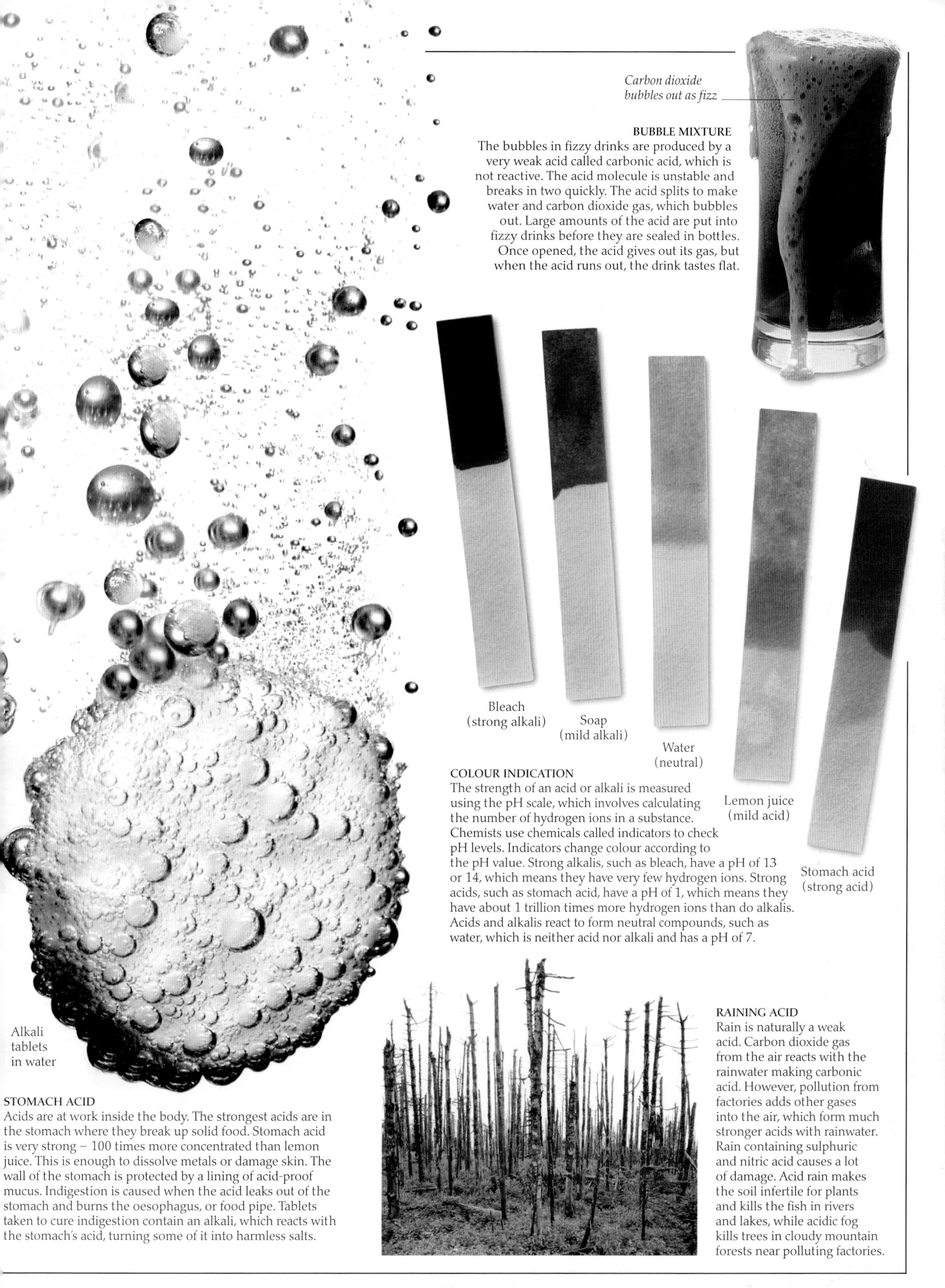

BUBBLE MIXTURE

The bubbles in fizzy drinks are produced by a very weak acid called carbonic acid, which is not reactive. The acid molecule is unstable and breaks in two quickly. The acid splits to make water and carbon dioxide gas, which bubbles out. Large amounts of the acid are put into fizzy drinks before they are sealed in bottles. Once opened, the acid gives out its gas, but when the acid runs out, the drink tastes flat.

COLOUR INDICATION

The strength of an acid or alkali is measured using the pH scale, which involves calculating the number of hydrogen ions in a substance. Chemists use chemicals called indicators to check pH levels. Indicators change colour according to the pH value. Strong alkalis, such as bleach, have a pH of 13 or 14, which means they have very few hydrogen ions. Strong acids, such as stomach acid, have a pH of 1, which means they have about 1 trillion times more hydrogen ions than do alkalis. Acids and alkalis react to form neutral compounds, such as water, which is neither acid nor alkali and has a pH of 7.

Alkali tablets in water

STOMACH ACID

Acids are at work inside the body. The strongest acids are in the stomach where they break up solid food. Stomach acid is very strong – 100 times more concentrated than lemon juice. This is enough to dissolve metals or damage skin. The wall of the stomach is protected by a lining of acid-proof mucus. Indigestion is caused when the acid leaks out of the stomach and burns the oesophagus, or food pipe. Tablets taken to cure indigestion contain an alkali, which reacts with the stomach's acid, turning some of it into harmless salts.

RAINING ACID

Rain is naturally a weak acid. Carbon dioxide gas from the air reacts with the rainwater making carbonic acid. However, pollution from factories adds other gases into the air, which form much stronger acids with rainwater. Rain containing sulphuric and nitric acid causes a lot of damage. Acid rain makes the soil infertile for plants and kills the fish in rivers and lakes, while acidic fog kills trees in cloudy mountain forests near polluting factories.

Getting electric

CHARGED GEM
The word "electric" comes from the Greek word for amber – *elektron*. Amber is a lump of ancient fossilized tree resin. Ancient Greeks found that when they rubbed amber with cloth, dust and hair clung to it. Rubbing a balloon produces the same effect. Scientists now understand that amber picks up an electric charge, and things stick to it so that the charge is balanced out.

ELECTRICITY IS ONE of the basic forces of nature. Every atom contains particles that carry electrical charge, which can be either positive (protons) or negative (electrons). Particles with the same kind of charge repel each other, while opposite charges attract. If charged particles cannot move about, a charge of static electricity can build up in one place. Charged particles can sometimes flow, however, as free electrons in a metal. They form an electric current. Electric wires carry these currents, but electricity can also be a flood of ions (charged atoms or molecules) moving through a liquid, such as the acid inside a battery. Electricity is useful because the power of the moving particles can be converted into heat, light, movement, and other handy forms of energy.

Glass pipes support the pile

Voltaic pile

Copper disc

Zinc disc

PILE OF ELECTRICITY
At first, scientists thought that electricity was created inside bodies as a mysterious "life force". However, in 1800, the Italian scientist Alessandro Volta invented an electric battery, which produced electricity using chemical reactions. The battery, known as a Voltaic pile, contained a pile of copper and zinc discs bathed in acid. As the acid and metals reacted, electrons moved through the pile. A wire running from the top of the pile to the bottom created a simple circuit for the electric current to travel around. Most of today's cylindrical batteries work in exactly the same way.

Similarly charged hair strands repel each other

Van de Graaff generator – a machine that produces a static electric charge

PUSHES AND PULLS
The girl touching this Van de Graaff generator is electrically charged. The charge is static and does not move so there is no dangerous current running through her. When objects have opposite charges, they attract each other, and balance out. However, objects with the same charge repel, or push away from each other. This repulsion is creating the spiky hairstyle. Each strand of hair has the same charge, and so is pushing away from all the others.

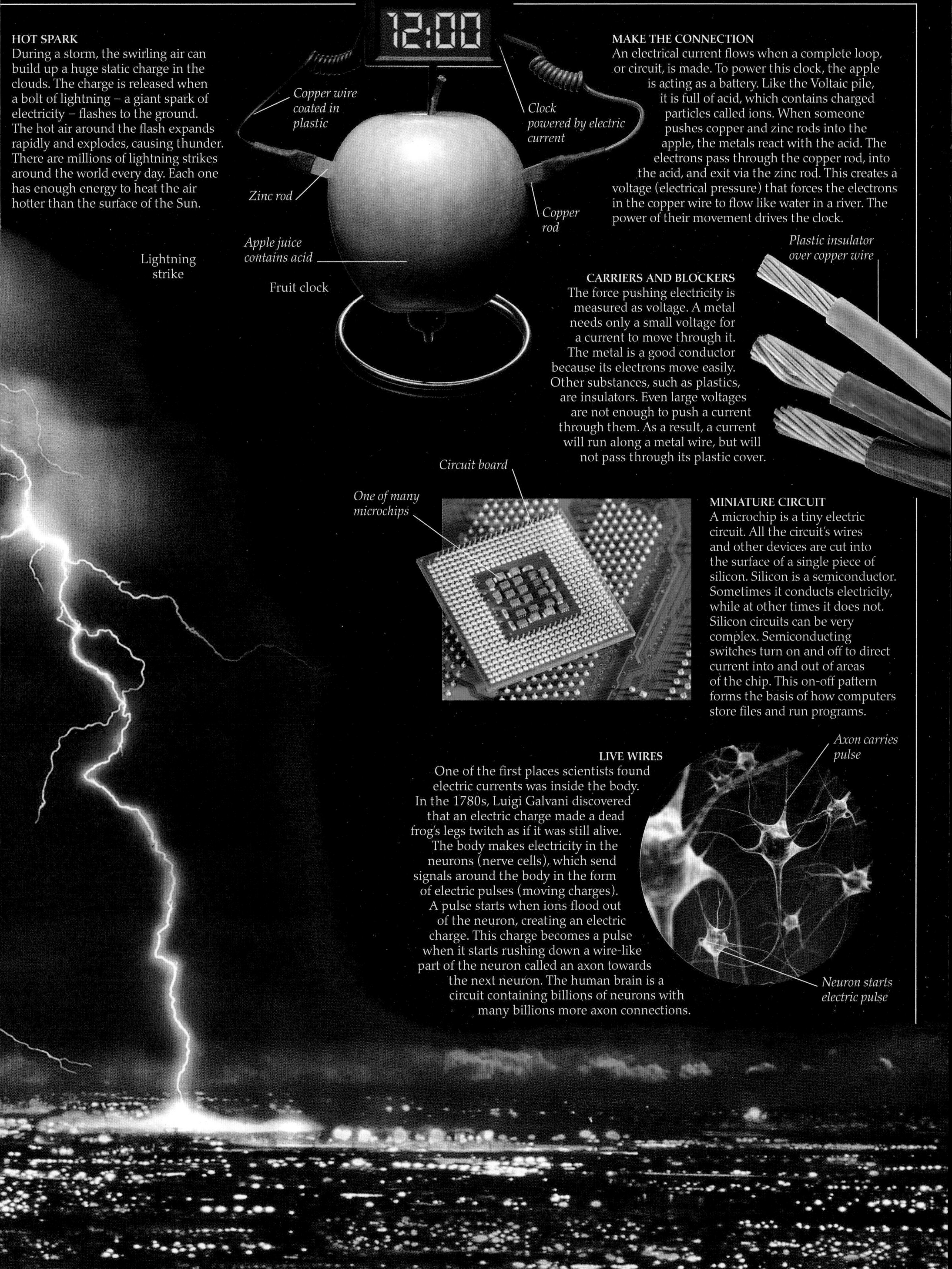

Lightning strike

Fruit clock

HOT SPARK

During a storm, the swirling air can build up a huge static charge in the clouds. The charge is released when a bolt of lightning – a giant spark of electricity – flashes to the ground. The hot air around the flash expands rapidly and explodes, causing thunder. There are millions of lightning strikes around the world every day. Each one has enough energy to heat the air hotter than the surface of the Sun.

MAKE THE CONNECTION

An electrical current flows when a complete loop, or circuit, is made. To power this clock, the apple is acting as a battery. Like the Voltaic pile, it is full of acid, which contains charged particles called ions. When someone pushes copper and zinc rods into the apple, the metals react with the acid. The electrons pass through the copper rod, into the acid, and exit via the zinc rod. This creates a voltage (electrical pressure) that forces the electrons in the copper wire to flow like water in a river. The power of their movement drives the clock.

CARRIERS AND BLOCKERS

The force pushing electricity is measured as voltage. A metal needs only a small voltage for a current to move through it. The metal is a good conductor because its electrons move easily. Other substances, such as plastics, are insulators. Even large voltages are not enough to push a current through them. As a result, a current will run along a metal wire, but will not pass through its plastic cover.

MINIATURE CIRCUIT

A microchip is a tiny electric circuit. All the circuit's wires and other devices are cut into the surface of a single piece of silicon. Silicon is a semiconductor. Sometimes it conducts electricity, while at other times it does not. Silicon circuits can be very complex. Semiconducting switches turn on and off to direct current into and out of areas of the chip. This on-off pattern forms the basis of how computers store files and run programs.

LIVE WIRES

One of the first places scientists found electric currents was inside the body. In the 1780s, Luigi Galvani discovered that an electric charge made a dead frog's legs twitch as if it was still alive. The body makes electricity in the neurons (nerve cells), which send signals around the body in the form of electric pulses (moving charges). A pulse starts when ions flood out of the neuron, creating an electric charge. This charge becomes a pulse when it starts rushing down a wire-like part of the neuron called an axon towards the next neuron. The human brain is a circuit containing billions of neurons with many billions more axon connections.

Magnetic attraction

THE WORD MAGNET comes from a region of Greece called Magnesia. People began making pure iron here around 3,000 years ago. Metal-workers found that some iron-rich rocks stuck to each other and clung to pieces of pure iron. These natural magnets – later named lodestones – were also found in India and China. Early people thought the pull of magnets was magical. It was only about 200 years ago that scientists began to explain it. They found that an invisible field of force – the magnetic field – surrounds a magnet and affects objects made of iron and steel in its vicinity. Iron and nickel are among a few metals that make good magnets, but magnetism has no effect on other metals, such as aluminium, tin, or copper.

Compass needle lines up with the loop of the force field

Iron filings follow shape of magnetic field

North pole of magnet

Pointed lodestone on Chinese compass

FINDING NORTH
The compass was probably invented in China about 1,800 years ago and people used it in rituals because they believed it had magical properties. Later, compasses began to be used for navigation. A compass contains a magnet that can freely swing in all directions. The small compass magnet is pulled on by a much larger one – Earth itself – and so it always points to the north. This Chinese compass is used in rituals to point to sacred places in the landscape.

INVISIBLE FORCE
Magnets are surrounded by an invisible force field that attracts magnetic substances such as iron. Magnets push and pull on other magnets. The force field loops around the magnet from one end, called the pole, to the other. The north pole of the magnet points to the north of Earth's magnetic field, and the south pole is at the other end. Opposite poles on different magnets attract each other, but like poles push each other away.

MAGNETIC PLANET
Earth's core is made of iron, the commonest magnetic metal. An inner core of solid iron spins within an outer layer of hot liquid iron. This creates an immense magnetic field that reaches far out into space. Every day the planet is blasted by solar wind – dangerous rays and particles from the Sun. The magnetic field blocks this deadly stream and funnels it down to Earth's poles. There, the solar wind smashes into the atmosphere, creating "aurorae" light shows – also called the Northern and Southern Lights.

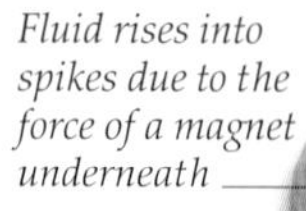

Fluid rises into spikes due to the force of a magnet underneath

LIQUID MAGNETS
Only solids can be magnetic, but scientists have found a way of making liquid behave as magnets. This liquid, called ferrofluid, is made of millions of tiny specks of magnetic iron mixed in an oil. A magnet held underneath affects the iron particles in the ferrofluid, making it flow into unusual shapes. Ferrofluids are used as lubricating oils inside machines that use moving magnets. Their magnetism holds them near the magnets.

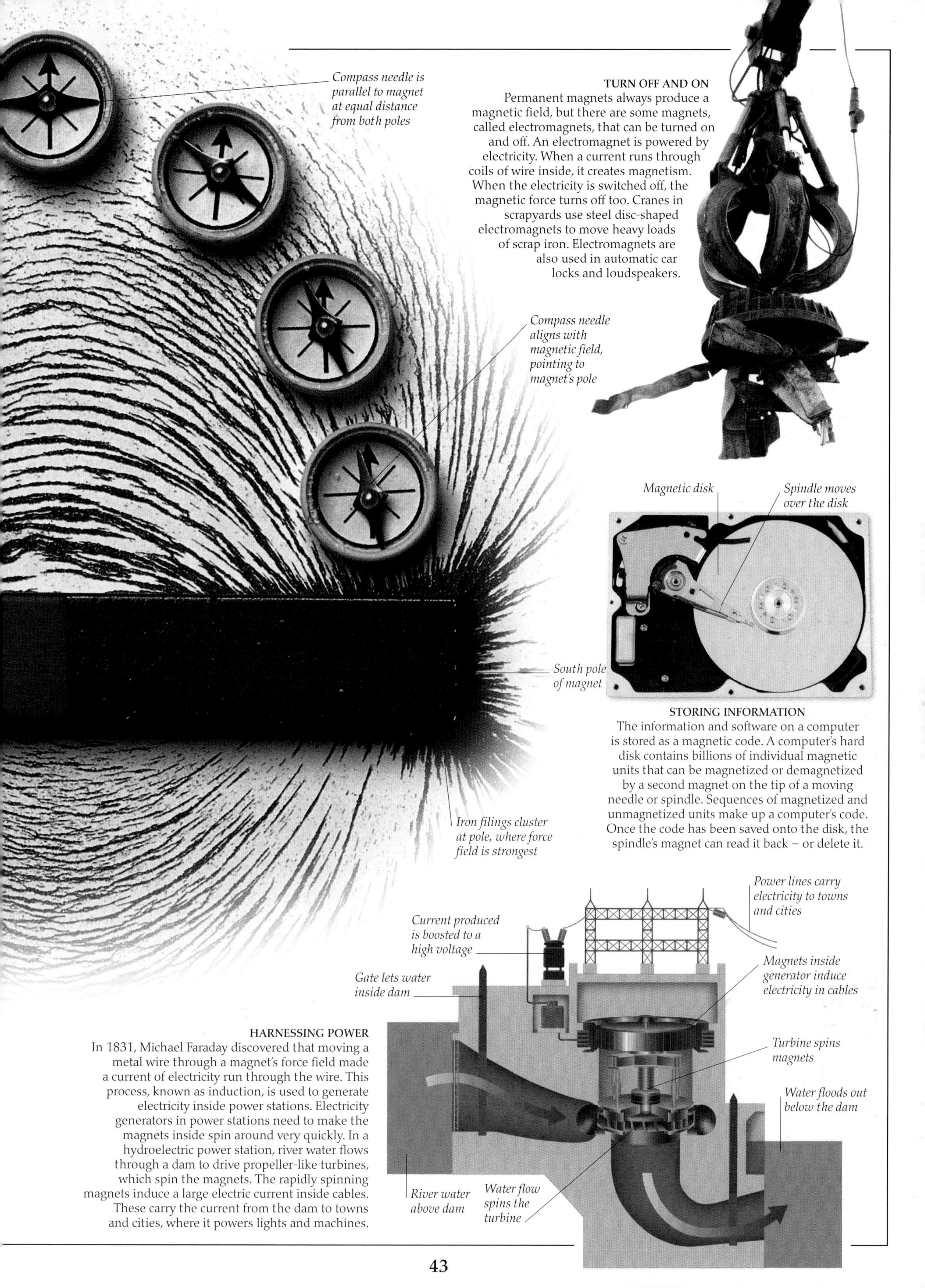

TURN OFF AND ON

Permanent magnets always produce a magnetic field, but there are some magnets, called electromagnets, that can be turned on and off. An electromagnet is powered by electricity. When a current runs through coils of wire inside, it creates magnetism. When the electricity is switched off, the magnetic force turns off too. Cranes in scrapyards use steel disc-shaped electromagnets to move heavy loads of scrap iron. Electromagnets are also used in automatic car locks and loudspeakers.

STORING INFORMATION

The information and software on a computer is stored as a magnetic code. A computer's hard disk contains billions of individual magnetic units that can be magnetized or demagnetized by a second magnet on the tip of a moving needle or spindle. Sequences of magnetized and unmagnetized units make up a computer's code. Once the code has been saved onto the disk, the spindle's magnet can read it back – or delete it.

HARNESSING POWER

In 1831, Michael Faraday discovered that moving a metal wire through a magnet's force field made a current of electricity run through the wire. This process, known as induction, is used to generate electricity inside power stations. Electricity generators in power stations need to make the magnets inside spin around very quickly. In a hydroelectric power station, river water flows through a dam to drive propeller-like turbines, which spin the magnets. The rapidly spinning magnets induce a large electric current inside cables. These carry the current from the dam to towns and cities, where it powers lights and machines.

Buried chemicals

Most of the fuels that people use come from petroleum. The word petroleum means rock oil, but it is really a mixture of chemicals made of hundreds of different solids, liquids, and gases. Petroleum chemicals are described as organic because they formed from the remains of ancient living things buried deep underground for millions of years. Most of the organic chemicals in petroleum are made from just carbon and hydrogen atoms and are known as hydrocarbons. A single carbon atom can form strong bonds with four other atoms. This allows carbon to form very complicated molecules made up of long chains and rings of atoms. No one knows how many organic compounds there are, but so far, scientists have recorded several million.

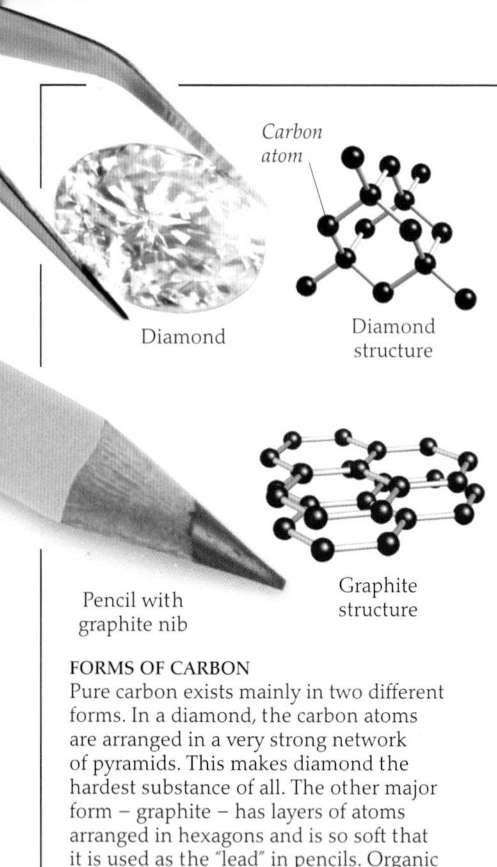

FORMS OF CARBON
Pure carbon exists mainly in two different forms. In a diamond, the carbon atoms are arranged in a very strong network of pyramids. This makes diamond the hardest substance of all. The other major form – graphite – has layers of atoms arranged in hexagons and is so soft that it is used as the "lead" in pencils. Organic molecules also contain similar hexagonal and pyramidal structures and can even form into complex ball shapes with 60 or more carbon atoms.

OIL FROM ROCK
Petroleum formed from dead plankton (tiny living organisms floating in the seas) that settled in thick layers of ooze on the sea bed millions of years ago. Mud and sand covered the ooze until it was buried deep within. Without any oxygen to react with it, the ooze gradually broke down into a sludge of hydrocarbons. A lot of the oil trickled to the surface forming tar pits or mixing into the ocean, but some was trapped in underground lakes of petroleum. People drill deep through the rocks in Earth's crust – sometimes many kilometres (miles) down – to tap into these hidden reservoirs.

HYDROCARBONS
The chemicals in petroleum are grouped according to how large their molecules are. Natural gas is mainly methane, the simplest hydrocarbon, which contains just one carbon atom. Thick tar chemicals contain more than 70 carbon atoms. The groups, or fractions, of petroleum separate when heated inside a tall tower. The smaller chemicals, such as those in petrol, float to the top, while the heavier chemicals do not rise as high.

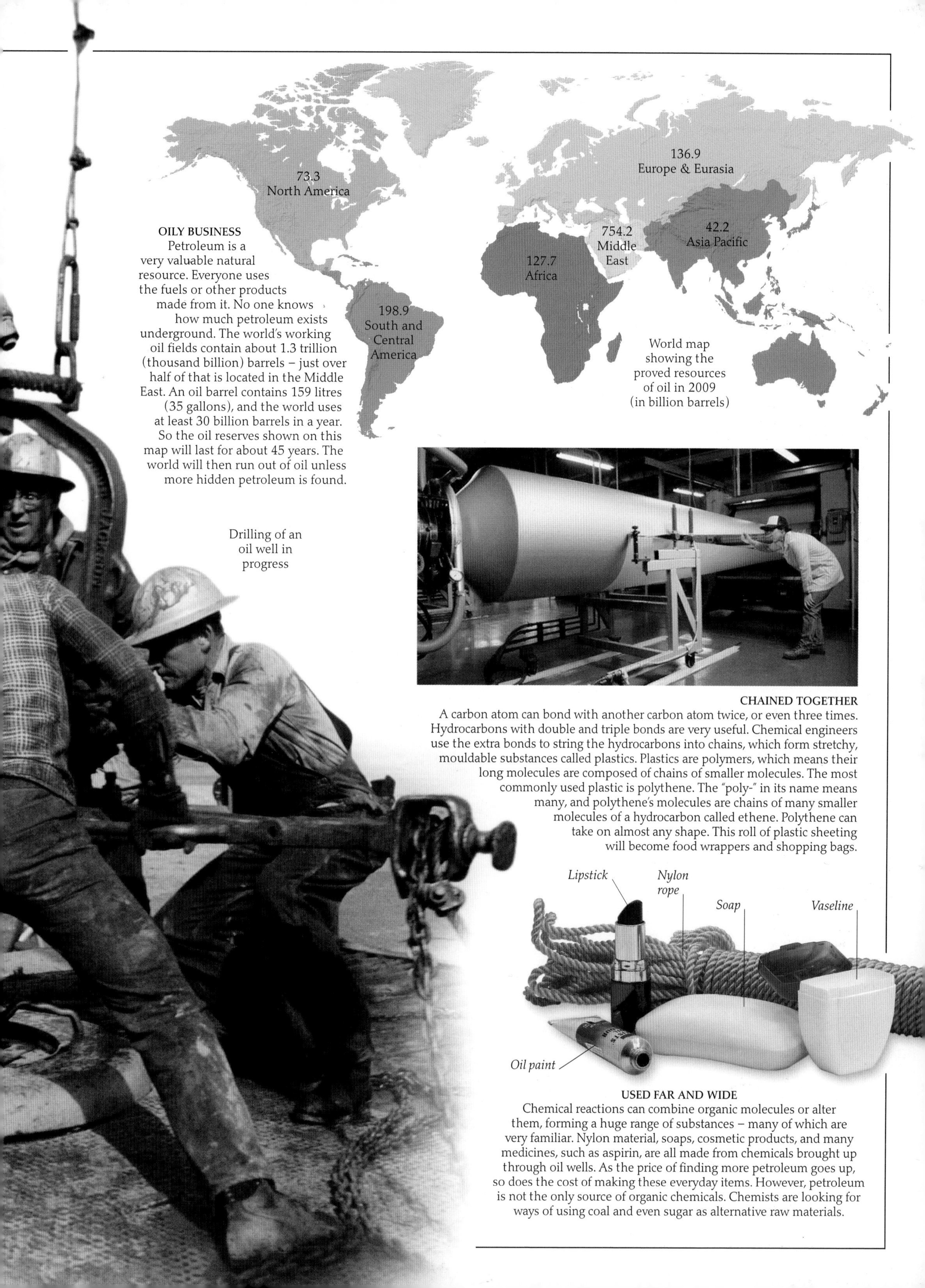

World map showing the proved resources of oil in 2009 (in billion barrels)

OILY BUSINESS

Petroleum is a very valuable natural resource. Everyone uses the fuels or other products made from it. No one knows how much petroleum exists underground. The world's working oil fields contain about 1.3 trillion (thousand billion) barrels – just over half of that is located in the Middle East. An oil barrel contains 159 litres (35 gallons), and the world uses at least 30 billion barrels in a year. So the oil reserves shown on this map will last for about 45 years. The world will then run out of oil unless more hidden petroleum is found.

Drilling of an oil well in progress

CHAINED TOGETHER

A carbon atom can bond with another carbon atom twice, or even three times. Hydrocarbons with double and triple bonds are very useful. Chemical engineers use the extra bonds to string the hydrocarbons into chains, which form stretchy, mouldable substances called plastics. Plastics are polymers, which means their long molecules are composed of chains of smaller molecules. The most commonly used plastic is polythene. The "poly-" in its name means many, and polythene's molecules are chains of many smaller molecules of a hydrocarbon called ethene. Polythene can take on almost any shape. This roll of plastic sheeting will become food wrappers and shopping bags.

USED FAR AND WIDE

Chemical reactions can combine organic molecules or alter them, forming a huge range of substances – many of which are very familiar. Nylon material, soaps, cosmetic products, and many medicines, such as aspirin, are all made from chemicals brought up through oil wells. As the price of finding more petroleum goes up, so does the cost of making these everyday items. However, petroleum is not the only source of organic chemicals. Chemists are looking for ways of using coal and even sugar as alternative raw materials.

Atomic energy

SCIENTISTS HAVE LEARNED how to harness energy from atoms in two ways. The first is nuclear fission which breaks an atom's nucleus apart. This is what happens inside nuclear power stations. The fuels used are radioactive elements (see page 25), such as uranium, which have large, unstable atoms. One gram (0.04 oz) of nuclear fuel holds 1 million times more energy than the same amount of coal. The second method is nuclear fusion, which smashes small atoms together so hard that they fuse, or join, making larger atoms. Both processes can turn into runaway chain reactions used for nuclear bombs, but to achieve an explosion, careful engineering is needed.

SPLITTING THE ATOM
Enrico Fermi was the first scientist to split the atom. He built a nuclear reactor in Chicago, US, in 1942, to try to control the fission process. Fermi did not understand just how dangerous radioactive elements are, and in 1954, he died from an illness caused by working with nuclear fuels for too long. Today, radioactive materials are stored safely.

Unstable nucleus
Ejected neutron
Proton
Nucleus splits in two
Heat, light, and other radiation released
Smaller nucleus formed
Ejected neutron
Larger, unstable nucleus ready to split

CHAIN REACTION
Radioactive elements are unstable because their atoms are big and unwieldy. The protons and neutrons cling together only weakly in the atom's nucleus. In fact, uranium's nuclei spontaneously lose protons and neutrons, giving off radiation. When a lot of uranium is put together, the ejected neutrons start a fission chain reaction. They collide with other nuclei, causing each of them to split in two, releasing energy and ejecting more neutrons that bombard yet more nuclei. This behaviour allows people to use uranium as nuclear fuel in reactors. Nuclear reactors regulate the number of neutrons released in the reaction, so the energy can be generated slowly and safely.

ATOMIC TESTING
In atomic bombs, nuclear fission occurs at runaway speed, resulting in an enormous explosion. In 1945, two atomic bombs were dropped by American warplanes on Japan at the end of World War II. Since then, bomb testing is only done in uninhabited areas, such as this site in the Nevada desert, US. The bomb produces a mushroom-shaped cloud and blasts out a massive amount of radioactive dust, which is as dangerous as the explosion.

Radioactive dust

SOLAR POWER

All stars generate energy through nuclear fusion. The Sun, like every other star, is mainly a ball of hydrogen atoms. At the heart of the Sun, the weight of all the gas pushing down is 250 billion times higher than the pressure of air on Earth. This is strong enough to trigger the fusion of hydrogen atoms. All the light, heat, and ultraviolet radiation that comes from the Sun is released by these fusion reactions at its core, but it takes years for the energy to find its way to the surface. From there, the light shines out into space and reaches Earth in just 8 minutes!

Mushroom cloud

Ultraviolet image of the Sun

Deuterium nucleus

Ejected neutron

Proton

Tritium nucleus

Fusion releases more energy than fission

Helium nucleus

TWO BECOMES ONE

Most fusion reactions use hydrogen, the simplest element with the smallest atoms. Most hydrogen atoms contain no neutrons, but some rare forms of hydrogen have neutrons in their nuclei. Hydrogen atoms with one neutron are known as deuterium, and atoms with two neutrons are called tritium. When a deuterium nucleus is made to collide with a tritium nucleus, they fuse into a single helium nucleus, which is a different element. The atoms of all heavier elements found on Earth were also made by fusion, inside stars.

Scientist preparing the insides of a fusion reactor

RING OF POWER

Scientists are building the first fusion power stations that could solve a lot of energy problems. Fusion reactors are fuelled with hydrogen, the most common element in the Universe. These reactors have a giant, doughnut-like vessel, which carries hydrogen plasma (see page 33) heated to millions of degrees. A magnetic field keeps the plasma running around in rings inside. Two streams of hydrogen atoms whizz around inside in opposite directions before smashing into each other so fast that they fuse and release energy. Unlike other power plants, nuclear fusion does not create any waste, and one fusion plant could replace four fission plants.

The chemistry of life

Stamp showing urea

ACCIDENTAL DISCOVERY
Scientists thought that the chemicals inside the body contained a mysterious "life force", and it was impossible to create these chemicals in a laboratory. In 1828, German chemist Friedrich Wöhler accidentally created urea – a simple chemical in urine – in his laboratory. This suggested that urea and other substances are created by chemical reactions inside living bodies. The discovery was commemorated by a German stamp.

ALL LIVING THINGS, from bacteria to humans, are made up mainly of salty water, but every organism contains about 60 elements, combined into thousands of different compounds called biochemicals. Most biochemicals are made primarily from hydrogen, oxygen, and carbon, but some also include atoms of minerals such as copper, zinc, and iodine. This makes biochemicals similar in composition to the compounds found in petroleum, which is the remains of living things. Biochemicals generally belong to one of three groups – carbohydrates, lipids, and proteins. These groups act as the building blocks and sources of fuel of all living things.

Oxygen atom – one of two at the end of every amino acid making up the albumin protein molecule

Energy released as light and heat

LIPIDS
There are two kinds of lipids. Plants have liquid lipids, known as oils, such as olive oil. Animal bodies contain solid lipids called fats. All lipid molecules have a similar structure – three chain-shaped acids that dangle from a top section, making the molecule look like chemical jellyfish. The acid "tentacles" of fat molecules are packed with hydrogen atoms, which makes them stick to each other, so fats are sticky solids. Oils have less hydrogen in their molecules, so they do not cling together as much.

Olive oil

Olive

POWER PROTEIN
Proteins are chemicals that run the body like tiny molecular machines. A protein is a chain of hundreds of smaller chemicals called amino acids, of which there are 21 natural types. Every protein chain has a certain shape determined by its amino acid sequence. The shape is crucial to its function in the body. Albumin proteins are made in the liver and then added to the blood, where they draw water into the blood vessels to stop the vessels from drying out. Albumin is also a major ingredient in egg white. Proteins also work as enzymes, controlling most chemical reactions in the body.

Burning sugar

Cell storing fat

SOLID FUEL
Carbohydrates include sugars, starch, and cellulose. Sugars are the simplest carbohydrates and living bodies use them as fuel. Burning sugar releases energy. The body does not set them on fire, but processes them in energy-releasing chemical reactions. Starch is made of many sugar molecules and acts as a fuel store. It is the main ingredient of bread. Cellulose forms tough supports in plant tissues.

FATTEN UP
Fat is packed with energy, so some animals use it as a long-term store of food. When an animal has plenty to eat, its body saves some of the food as fat, which is stored inside cells around the body. When there is a severe shortage of food – such as in the middle of winter – the fat supply keeps the animal from starving. The human body also stores fats, but too much fat in the body damages the heart and obstructs the blood supply.

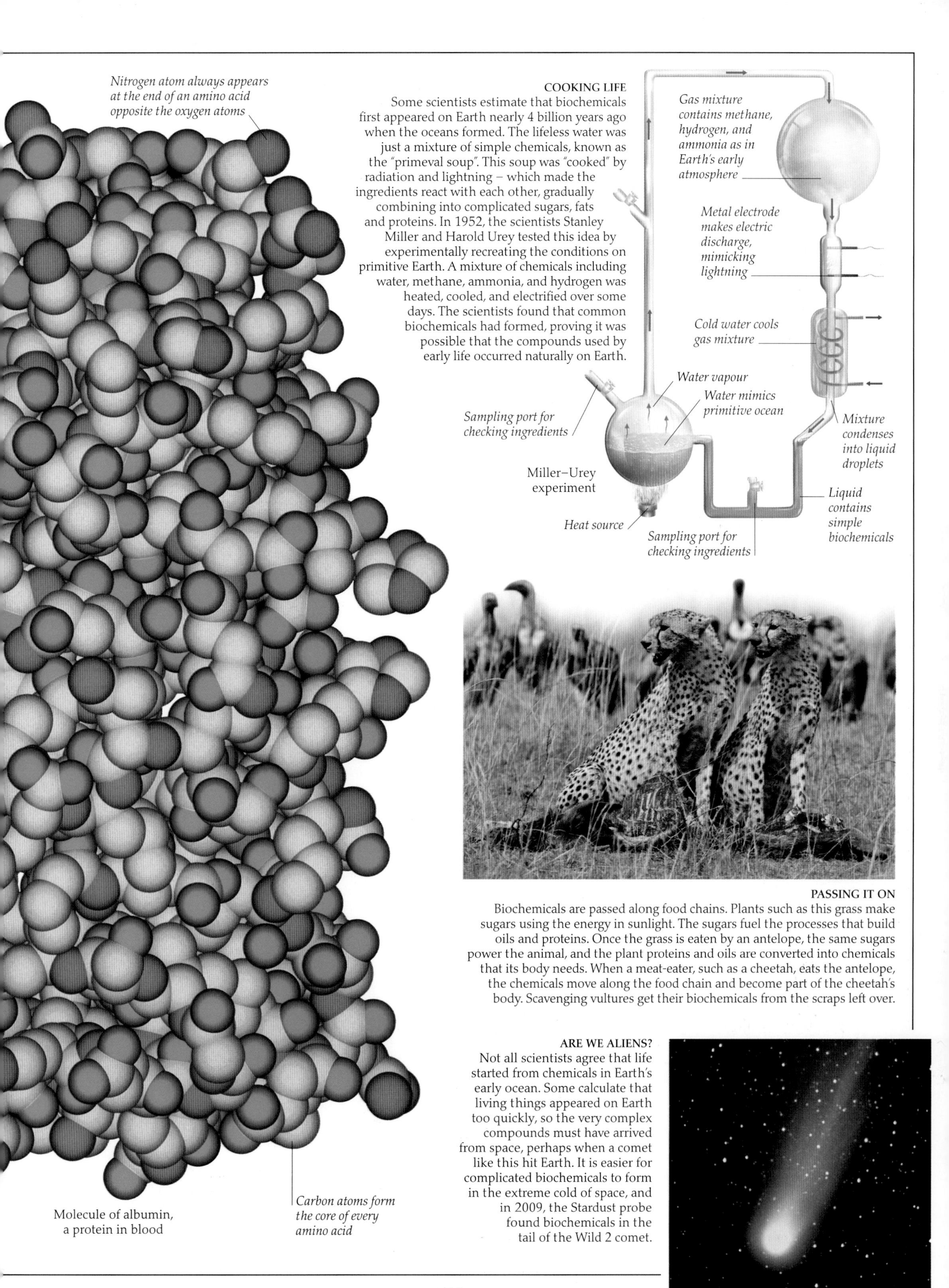

Molecule of albumin, a protein in blood

Miller–Urey experiment

COOKING LIFE

Some scientists estimate that biochemicals first appeared on Earth nearly 4 billion years ago when the oceans formed. The lifeless water was just a mixture of simple chemicals, known as the "primeval soup". This soup was "cooked" by radiation and lightning – which made the ingredients react with each other, gradually combining into complicated sugars, fats and proteins. In 1952, the scientists Stanley Miller and Harold Urey tested this idea by experimentally recreating the conditions on primitive Earth. A mixture of chemicals including water, methane, ammonia, and hydrogen was heated, cooled, and electrified over some days. The scientists found that common biochemicals had formed, proving it was possible that the compounds used by early life occurred naturally on Earth.

PASSING IT ON

Biochemicals are passed along food chains. Plants such as this grass make sugars using the energy in sunlight. The sugars fuel the processes that build oils and proteins. Once the grass is eaten by an antelope, the same sugars power the animal, and the plant proteins and oils are converted into chemicals that its body needs. When a meat-eater, such as a cheetah, eats the antelope, the chemicals move along the food chain and become part of the cheetah's body. Scavenging vultures get their biochemicals from the scraps left over.

ARE WE ALIENS?

Not all scientists agree that life started from chemicals in Earth's early ocean. Some calculate that living things appeared on Earth too quickly, so the very complex compounds must have arrived from space, perhaps when a comet like this hit Earth. It is easier for complicated biochemicals to form in the extreme cold of space, and in 2009, the Stardust probe found biochemicals in the tail of the Wild 2 comet.

The double helix

Single cell of an animal's body

Cell nucleus, the control centre containing the DNA

THE BODY OF A LIVING THING is built from units called cells and grows by following a set of instructions in its cells, held in batches called genes. The genes of every living thing are stored as a code on a chemical called deoxyribonucleic acid (DNA). Every cell has its own complete set of DNA. Scientists working more than 100 years ago understood that every life form inherits genes from its parents, so children grow up to look like their mothers and fathers. However, no one knew how the instructions were passed on until the 1950s, when scientists unravelled the mysteries of the DNA code. The field of science that looks at how DNA works and how genes are passed on is called genetics.

Chromosome is mainly a tightly packed DNA molecule

TWISTED CHEMICAL
The DNA molecule is a polymer – a chain made of thousands of smaller molecules. These smaller units include a sugar called ribose and one of four different nitrogen-containing molecules called bases. In 1953, scientists Francis Crick and James Watson built a model of how these units fitted together. They found that DNA was shaped like a twisted ladder, or a "double helix". The two sides of the helix, or spiral, were made from ribbons of ribose molecules, while pairs of the bases formed the "rungs" across the middle.

DNA strand is coiled up several times to pack into a chromosome

Scaffold made of proteins called histones allows effective coiling of DNA

LADDER CODE
The DNA code is stored in the rungs of a DNA molecule, which are made of the compounds known as bases. The four bases are called adenine, thymine, guanine, and cytosine – simplified by scientists as A, T, G, and C. In each DNA molecule, A always combines with T, and G pairs with C. The sequence of bases, which runs along the core of the DNA molecule, is always different and creates a code. Human DNA has 3 billion base pairs. Laid end to end it would be 3 m (9.8 ft) long, but far too thin to see. The long DNA strands are coiled away in compact structures called chromosomes. These are found in a store called the nucleus at the heart of each cell in all living things.

FROM GENE TO PROTEIN
DNA's code forms instructions for making useful proteins (see page 48). The instructions are carried on portions of the DNA strand called genes, with every gene making a different protein type. The ATGC code tells the cell the precise order of amino acids needed to build the protein. When the cell needs a protein, it unzips the DNA double helix, breaking apart the base pairs and creating a single string of bases. The cell then reads, copies, and translates this sequence into the protein it needs.

Opposite base on RNA

Base on DNA

RNA copy of DNA code

Second strand of DNA is not read

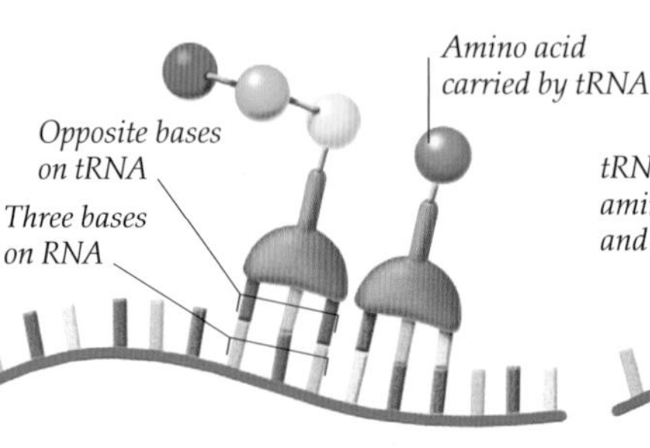

Amino acid chain grows into protein

tRNA releases amino acid and moves off

1 COPYING THE CODE
When the DNA is unzipped, its base sequence is exposed. Fragments of a related molecule called RNA float inside the cell. Each fragment contains one base. These pieces bond with the exposed DNA bases according to the A-T, G-C rules. An RNA copy of the DNA sequence builds up.

2 TRANSLATION
The RNA code (mRNA) is next translated into a protein using another type of RNA called transfer RNA (tRNA). Each tRNA molecule carries one amino acid, and has a three-base sequence unique to that amino acid. The bases on the tRNA match up with the base sequence on the mRNA.

3 PROTEIN MANUFACTURE
The sequence of bases on the mRNA means that the tRNA molecules line up in a set order, hauling their amino acids into the correct places to make the appropriate protein. The amino acids link together to make the protein coded by one gene. Each protein carries out a vital function in the body.

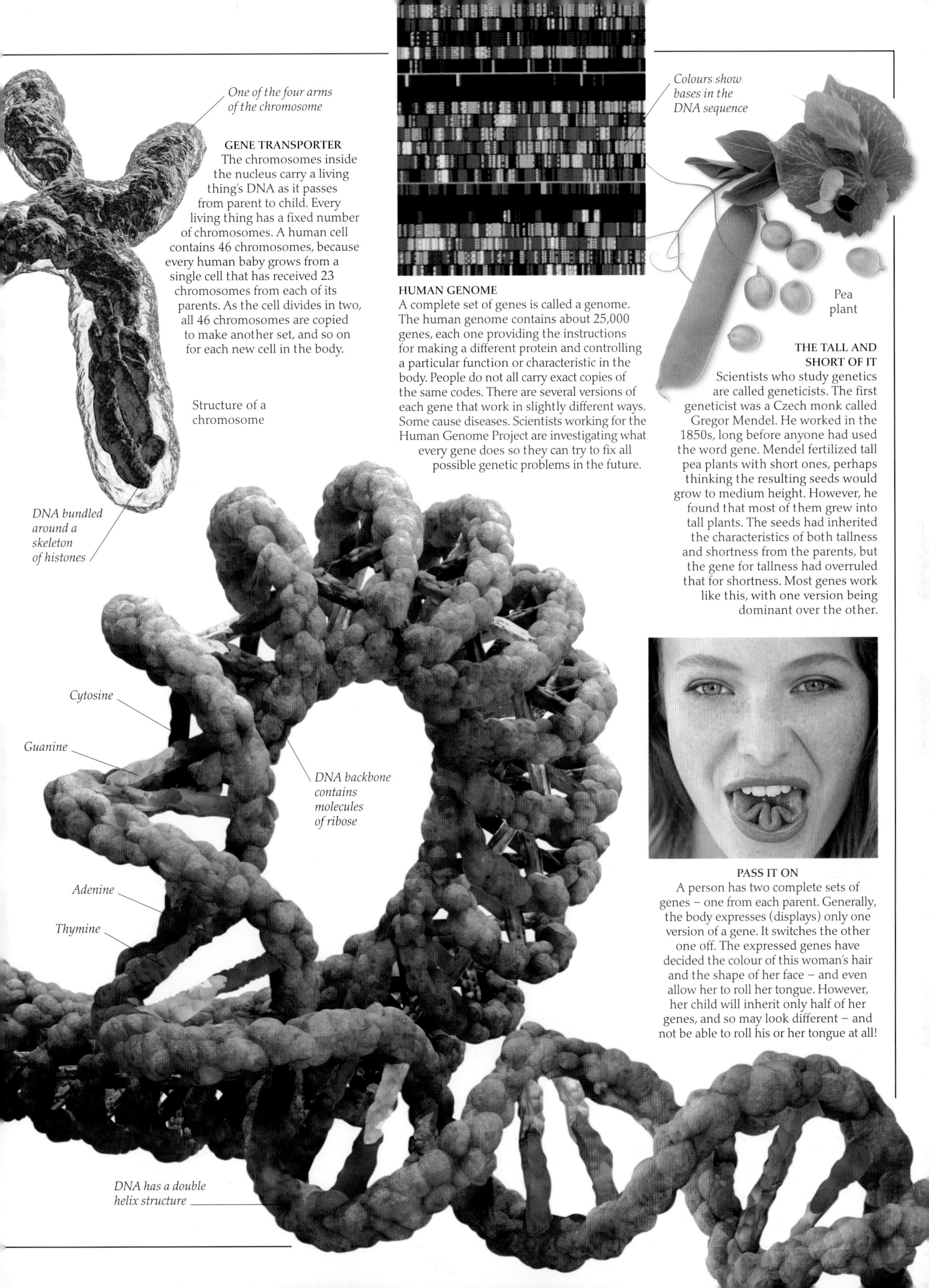

GENE TRANSPORTER
The chromosomes inside the nucleus carry a living thing's DNA as it passes from parent to child. Every living thing has a fixed number of chromosomes. A human cell contains 46 chromosomes, because every human baby grows from a single cell that has received 23 chromosomes from each of its parents. As the cell divides in two, all 46 chromosomes are copied to make another set, and so on for each new cell in the body.

HUMAN GENOME
A complete set of genes is called a genome. The human genome contains about 25,000 genes, each one providing the instructions for making a different protein and controlling a particular function or characteristic in the body. People do not all carry exact copies of the same codes. There are several versions of each gene that work in slightly different ways. Some cause diseases. Scientists working for the Human Genome Project are investigating what every gene does so they can try to fix all possible genetic problems in the future.

THE TALL AND SHORT OF IT
Scientists who study genetics are called geneticists. The first geneticist was a Czech monk called Gregor Mendel. He worked in the 1850s, long before anyone had used the word gene. Mendel fertilized tall pea plants with short ones, perhaps thinking the resulting seeds would grow to medium height. However, he found that most of them grew into tall plants. The seeds had inherited the characteristics of both tallness and shortness from the parents, but the gene for tallness had overruled that for shortness. Most genes work like this, with one version being dominant over the other.

PASS IT ON
A person has two complete sets of genes – one from each parent. Generally, the body expresses (displays) only one version of a gene. It switches the other one off. The expressed genes have decided the colour of this woman's hair and the shape of her face – and even allow her to roll her tongue. However, her child will inherit only half of her genes, and so may look different – and not be able to roll his or her tongue at all!

Evolution

EACH CELL OF EVERY LIVING ORGANISM carries its coded instructions in the form of DNA. When a cell divides, it copies the DNA, but mistakes often occur, altering the DNA code. These mistakes are called genetic mutations, and most are harmless. However, some can cause major problems, and even death. Occasionally, mutations can make an organism better at surviving and reproducing than other members of their species (a group of similar organisms that breed with each other). These organisms pass on their helpful mutation to some of their offspring, which become better adapted for survival. This process of change is called evolution. Over millions of years, many small but helpful mutations have allowed species to change, or evolve, into the plants, animals, and other forms of life that now populate Earth.

GETTING FIT
Charles Darwin was the scientist who worked out how evolution alters the way living things look and live. Darwin realized that some members of a species were "fitter" – better at surviving and breeding – than others. These animals passed this fitness to their offspring, which were strong, while the weaker animals died out. Darwin said that this process of "natural selection" could slowly change a group of animals into something new.

NATURAL SELECTION
Smoke from factories in the 19th century changed the environment around towns in Britain, creating an opportunity to see natural selection in action. Peppered moths were named after their pale speckled wings that camouflaged them on tree trunks covered in fungus and moss. The smoke darkened the bark, ruining the disguises of most speckled moths and making them easy prey for birds. A few moths had brown wings, which allowed more of them to survive. Through natural selection, British peppered moths gradually became darker.

Pale, speckled form of peppered moth

Intermediate form of peppered moth

Dark form of peppered moth

SPECIES FORMATION
New species form when groups from one species become isolated and natural selection changes them in different ways. This happened when a group of Asian macaque monkeys became cut off from the rest of their species on the islands of Japan. The two groups – one on the Asian mainland and the other in Japan – evolved into separate species as they adapted to different conditions. The Japanese animals became the snow monkey species, as they evolved to survive Japan's freezing winters. Unlike their relatives in warmer parts of Asia, snow monkeys grow thick fur to keep out the cold and migrate south to avoid the worst weather.

EVOLUTION
Every form of life on Earth today has evolved over millions of years from ancestors that looked very different. For example, elephants evolved from an ancestor called *Moeritherium*, which lived 37 million years ago (mya). *Moeritherium* lived in swamps and looked more like a hippopotamus than an elephant. As the environment changed, the species evolved through natural selection into other creatures that could survive better in the new environment.

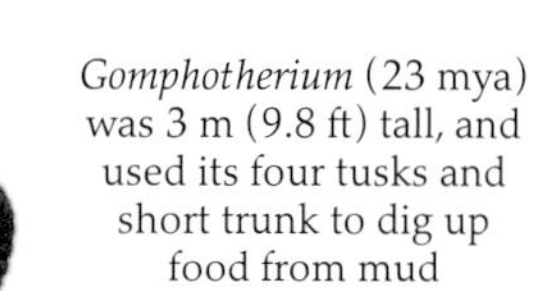

Gomphotherium (23 mya) was 3 m (9.8 ft) tall, and used its four tusks and short trunk to dig up food from mud

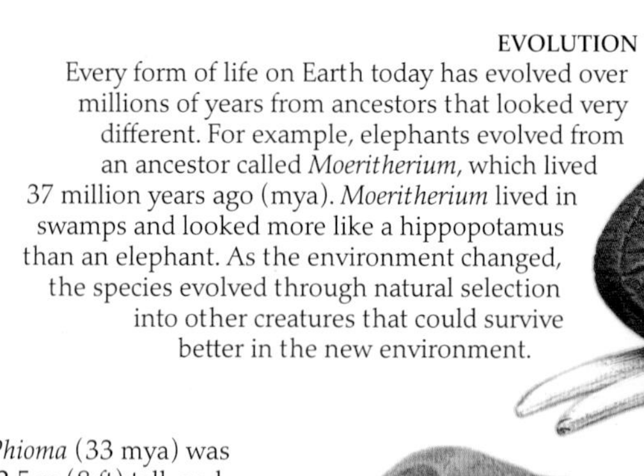

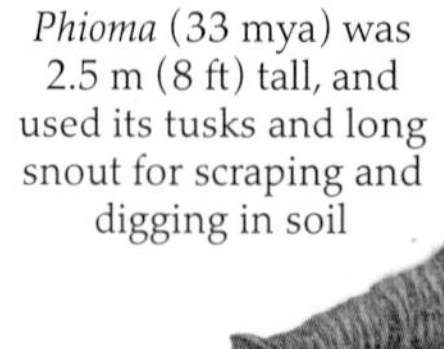

Phioma (33 mya) was 2.5 m (8 ft) tall, and used its tusks and long snout for scraping and digging in soil

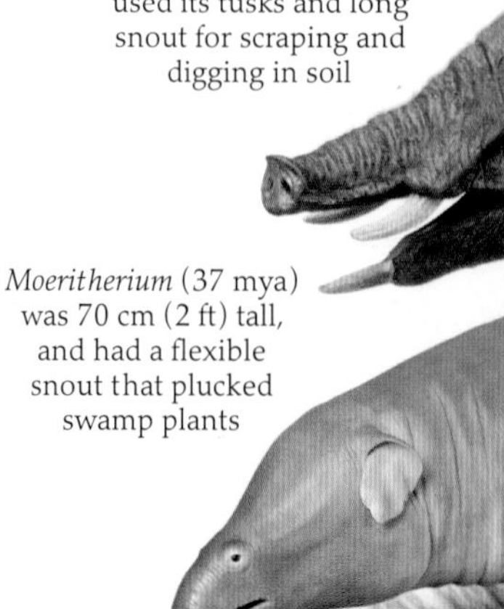

Moeritherium (37 mya) was 70 cm (2 ft) tall, and had a flexible snout that plucked swamp plants

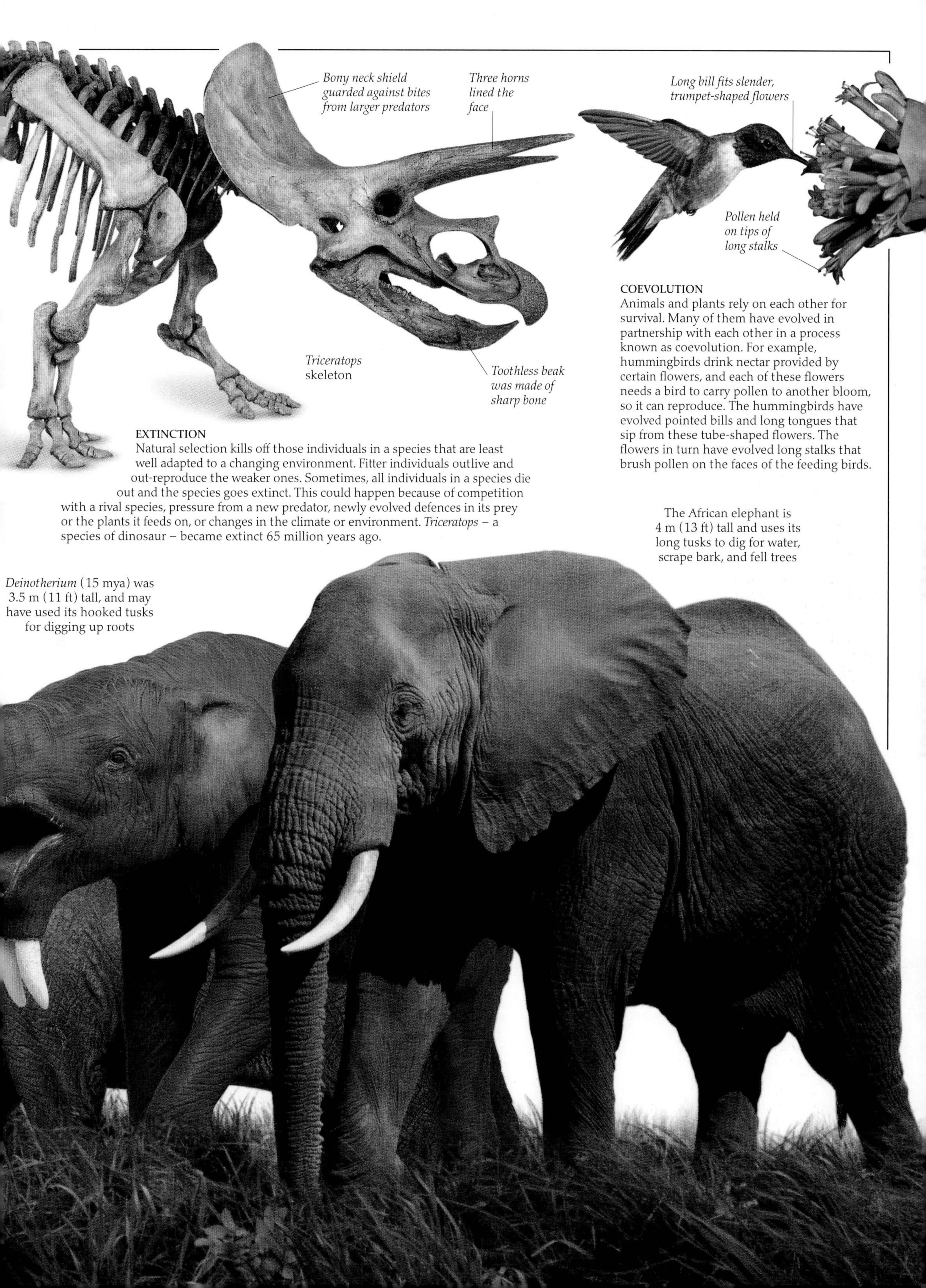

Triceratops skeleton

EXTINCTION

Natural selection kills off those individuals in a species that are least well adapted to a changing environment. Fitter individuals outlive and out-reproduce the weaker ones. Sometimes, all individuals in a species die out and the species goes extinct. This could happen because of competition with a rival species, pressure from a new predator, newly evolved defences in its prey or the plants it feeds on, or changes in the climate or environment. *Triceratops* – a species of dinosaur – became extinct 65 million years ago.

COEVOLUTION

Animals and plants rely on each other for survival. Many of them have evolved in partnership with each other in a process known as coevolution. For example, hummingbirds drink nectar provided by certain flowers, and each of these flowers needs a bird to carry pollen to another bloom, so it can reproduce. The hummingbirds have evolved pointed bills and long tongues that sip from these tube-shaped flowers. The flowers in turn have evolved long stalks that brush pollen on the faces of the feeding birds.

The African elephant is 4 m (13 ft) tall and uses its long tusks to dig for water, scrape bark, and fell trees

Deinotherium (15 mya) was 3.5 m (11 ft) tall, and may have used its hooked tusks for digging up roots

The cell

CELLS ARE THE BASIC UNITS that make up a living body. An adult human body contains around 100 trillion cells, while simple worms may have just a few thousand. Some life forms, such as amoebae and bacteria, have a body made from just a single cell. In larger organisms, body cells are not identical because they are specialized to do different jobs. Nerve cells are shaped like long wires and carry messages, while bone cells are encased in solid minerals that make the skeleton hard. Whatever their shape and size, all cells are surrounded by a flimsy envelope called the cell membrane, made of fat molecules. Inside the cell is a fluid called the cytoplasm. The cell membrane controls the chemicals moving in and out of the cytoplasm, making the cell a tiny world of its own.

Vacuole is a storage sac that releases substances such as waste by merging with the cell membrane

Protein microtubule hauls objects around the cell

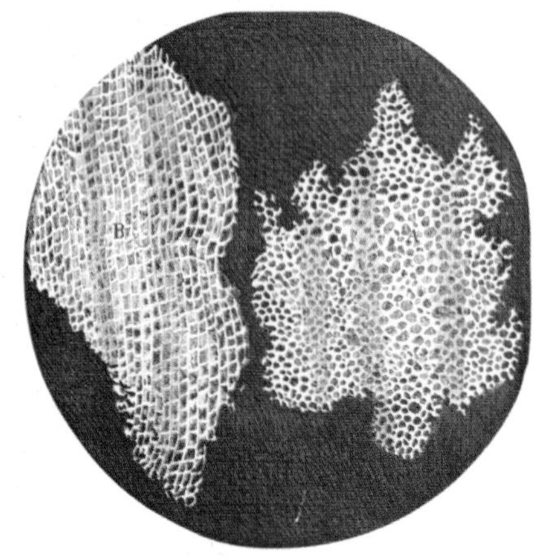

TINY SPACE
Scientist Robert Hooke came up with the name "cell" in the 17th century. He was one of the first people to use a microscope to look closely at animals and plants. He made detailed drawings of what he saw and thought that the little units that seemed to make up all plants and animals looked like the tiny rooms – or cells – used by monks.

INNER WORKINGS
Cells are a lot more than bags of watery liquid. The cytoplasm in cells is filled with smaller structures made from membranes folded into tubes and sacs. These are called organelles and form the machinery that keeps a cell alive and doing its job. The largest organelle is the nucleus, where DNA is stored. Others include the endoplasmic reticulum that makes proteins and lipids (see pages 48–49) and the lysosomes that contain powerful, germ-destroying enzymes (specialized proteins). Mitochondria are the cell's engines, releasing energy from sugars to power the cell (see page 57). Scientists think that cells packed full of organelles, each performing a specific function, evolved when a set of much smaller cells – such as bacteria – began to work together as a team.

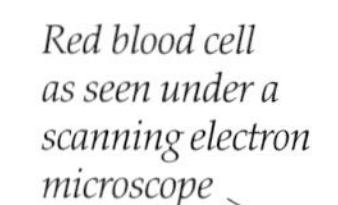

Red blood cell as seen under a scanning electron microscope

Cytoplasm is mainly water with minerals dissolved in it

TOO TINY
Most cells in the body are too small to be seen without a microscope. About 30 cells could be lined up across a full stop. Scientists use light microscopes, which create an image by focusing light with lenses, to study cells. However, the details inside the cells are too small for light to pick out. So electron microscopes, which use a beam of electrons, are needed to analyse a cell's interior. To look at the 3-D structure of cells, such as these red blood cells, biologists use a specialized microscope called a scanning electron microscope.

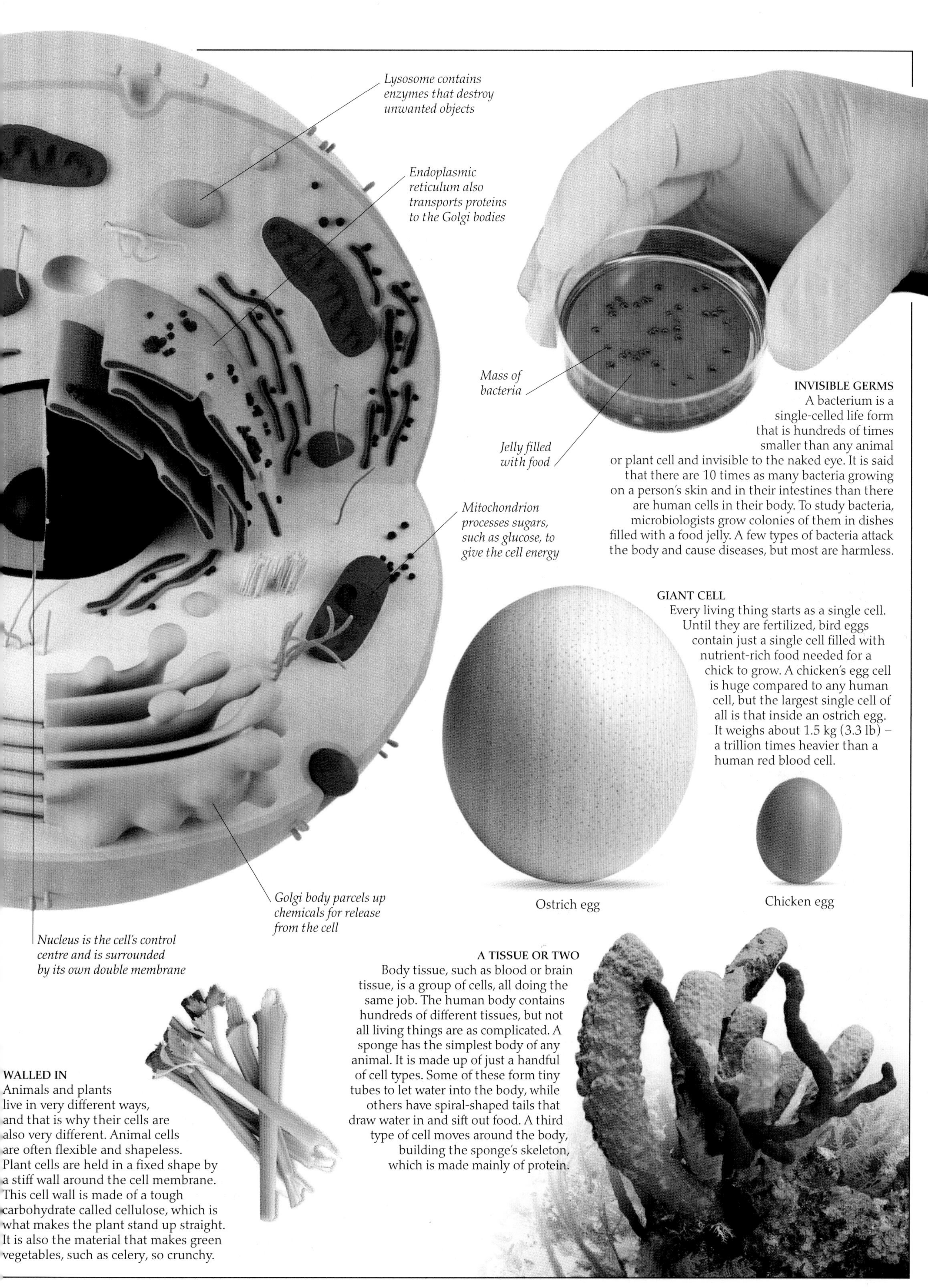

INVISIBLE GERMS

A bacterium is a single-celled life form that is hundreds of times smaller than any animal or plant cell and invisible to the naked eye. It is said that there are 10 times as many bacteria growing on a person's skin and in their intestines than there are human cells in their body. To study bacteria, microbiologists grow colonies of them in dishes filled with a food jelly. A few types of bacteria attack the body and cause diseases, but most are harmless.

GIANT CELL

Every living thing starts as a single cell. Until they are fertilized, bird eggs contain just a single cell filled with nutrient-rich food needed for a chick to grow. A chicken's egg cell is huge compared to any human cell, but the largest single cell of all is that inside an ostrich egg. It weighs about 1.5 kg (3.3 lb) – a trillion times heavier than a human red blood cell.

A TISSUE OR TWO

Body tissue, such as blood or brain tissue, is a group of cells, all doing the same job. The human body contains hundreds of different tissues, but not all living things are as complicated. A sponge has the simplest body of any animal. It is made up of just a handful of cell types. Some of these form tiny tubes to let water into the body, while others have spiral-shaped tails that draw water in and sift out food. A third type of cell moves around the body, building the sponge's skeleton, which is made mainly of protein.

WALLED IN

Animals and plants live in very different ways, and that is why their cells are also very different. Animal cells are often flexible and shapeless. Plant cells are held in a fixed shape by a stiff wall around the cell membrane. This cell wall is made of a tough carbohydrate called cellulose, which is what makes the plant stand up straight. It is also the material that makes green vegetables, such as celery, so crunchy.

Living energy

PLANTS AND ANIMALS ARE LIVING MACHINES. Every machine needs a power source, and living things are no different. An animal is fuelled by the food it eats. Food is full of useful fats and proteins for building new cells (see page 48), but it is the sugar in food that is used to power the body. Each of the animal's cells converts its supply of sugar fuel using a chemical process called respiration. Plant cells use sugar too, but unlike animals, plants create their own sugar using the energy coming directly from the Sun. This energy-capture process is called photosynthesis.

Sun's rays

Glucose released into plant's body

Water travels from roots to the leaves

Oxygen escapes from the plant's body fluids and flows out into the atmosphere

Carbon dioxide flows in from the atmosphere and dissolves in the plant's body fluids

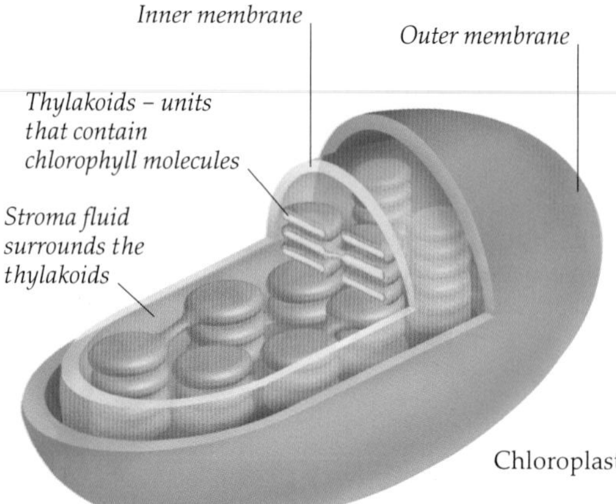

PLANT POWER

A plant is powered by the Sun. The leaves function like solar panels, which capture energy from sunlight. They absorb the red and blue colours in sunlight. They reflect the green light, which is why plants look green. The captured sunlight drives a process called photosynthesis, which means "making with light". Photosynthesis uses water from the plant's roots and combines it with carbon dioxide (CO_2) absorbed from the air. Six CO_2 molecules react with six molecules of water to make one molecule of glucose – the sugar used to power all life. The waste product of photosynthesis is oxygen, which is released into the air.

Inner membrane

Outer membrane

Thylakoids – units that contain chlorophyll molecules

Stroma fluid surrounds the thylakoids

Chloroplast

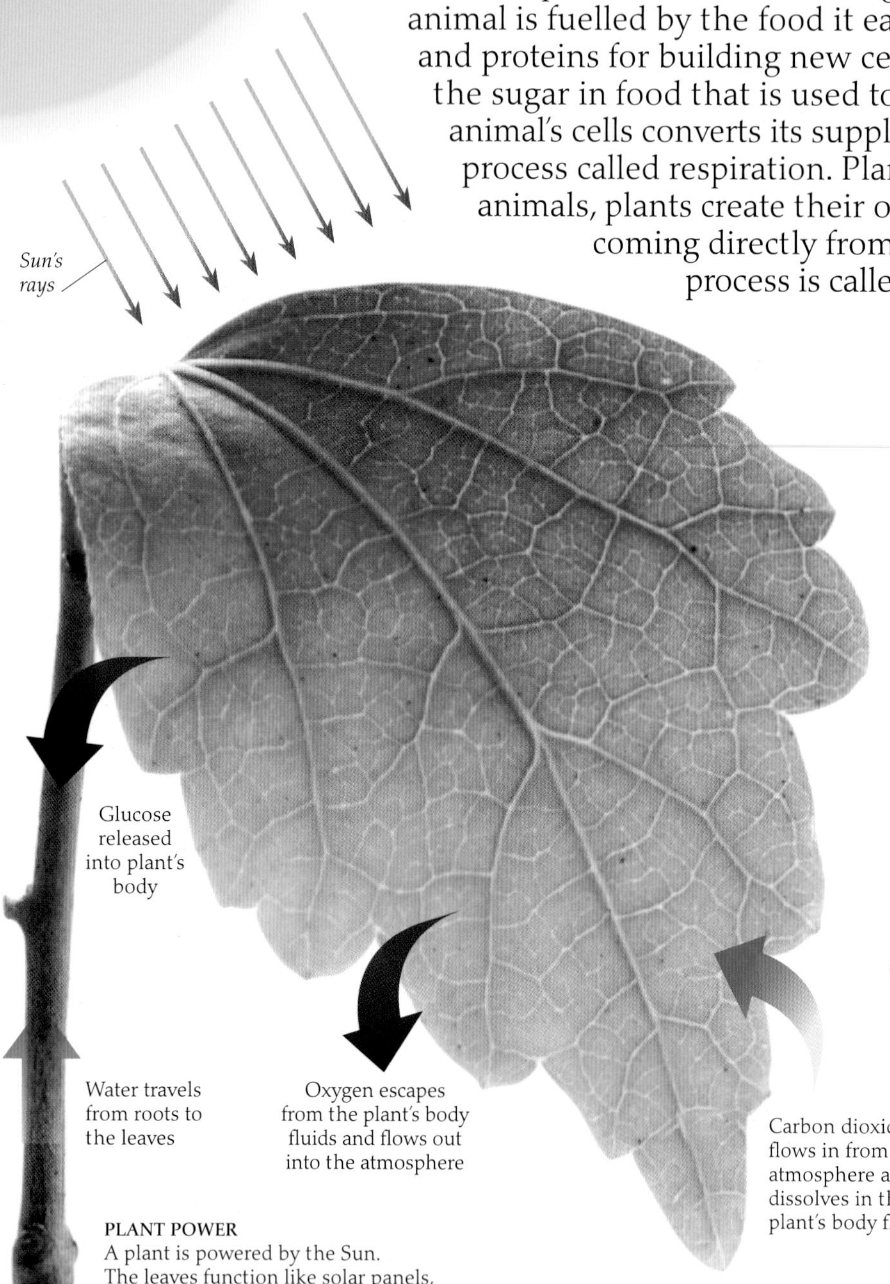

LIGHT TRAP

Photosynthesis takes place inside chloroplasts, which are found in most of the cells in a leaf – and in any other green parts of the plant. The green colouring comes from a chemical called chlorophyll. A chlorophyll molecule is sensitive to light. When sunlight hits one of these molecules, the energy in it triggers a complex series of chemical reactions that uses the captured energy to make energy-rich compounds. These pass into a thick fluid inside the chloroplast called the stroma. There, another chemical reaction uses the stored chemical energy to turn water and carbon dioxide into glucose.

Plant-eaters get only 10 per cent of energy captured by their plant food

ENERGY PYRAMID

Photosynthesis is the way the Sun's light energy fuels animal and plant life on Earth. Without photosynthesis in plants, animals could not exist. The glucose that powers an animal's body either comes straight from eating plants or from eating animals that feed on plants. In this way, the light energy trapped by plants and used to make glucose passes from one organism to another, forming a chain of food and energy. At each link in the chain, some of this energy is lost as it is used up and released as heat. As a result, fewer organisms survive at the top of a food chain than at the bottom.

Plant growth is powered by light from the Sun

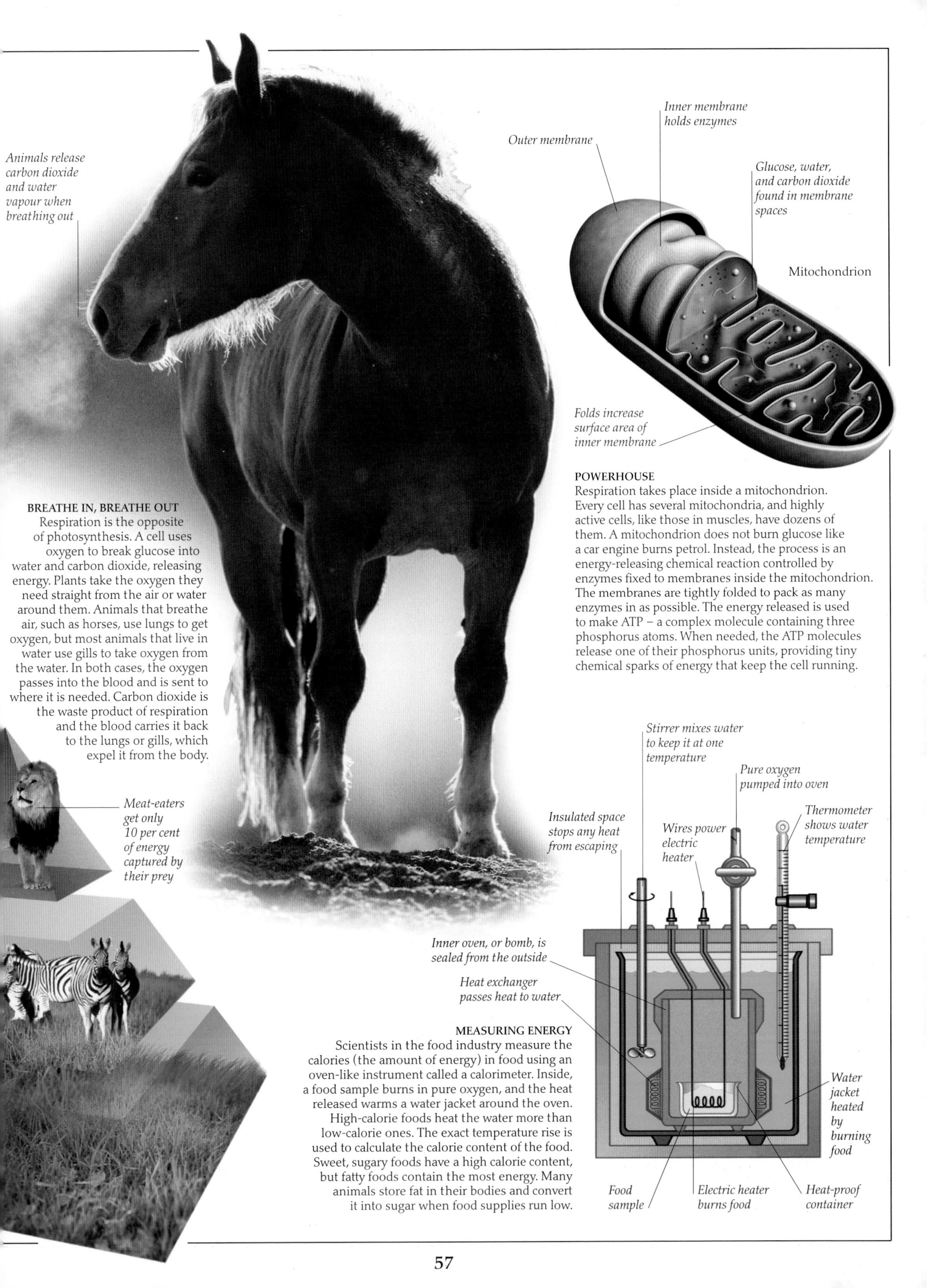

BREATHE IN, BREATHE OUT

Respiration is the opposite of photosynthesis. A cell uses oxygen to break glucose into water and carbon dioxide, releasing energy. Plants take the oxygen they need straight from the air or water around them. Animals that breathe air, such as horses, use lungs to get oxygen, but most animals that live in water use gills to take oxygen from the water. In both cases, the oxygen passes into the blood and is sent to where it is needed. Carbon dioxide is the waste product of respiration and the blood carries it back to the lungs or gills, which expel it from the body.

POWERHOUSE

Respiration takes place inside a mitochondrion. Every cell has several mitochondria, and highly active cells, like those in muscles, have dozens of them. A mitochondrion does not burn glucose like a car engine burns petrol. Instead, the process is an energy-releasing chemical reaction controlled by enzymes fixed to membranes inside the mitochondrion. The membranes are tightly folded to pack as many enzymes in as possible. The energy released is used to make ATP – a complex molecule containing three phosphorus atoms. When needed, the ATP molecules release one of their phosphorus units, providing tiny chemical sparks of energy that keep the cell running.

MEASURING ENERGY

Scientists in the food industry measure the calories (the amount of energy) in food using an oven-like instrument called a calorimeter. Inside, a food sample burns in pure oxygen, and the heat released warms a water jacket around the oven. High-calorie foods heat the water more than low-calorie ones. The exact temperature rise is used to calculate the calorie content of the food. Sweet, sugary foods have a high calorie content, but fatty foods contain the most energy. Many animals store fat in their bodies and convert it into sugar when food supplies run low.

Time and space

ONE OF THE GOALS OF SCIENCE is to find laws of nature that explain the way the Universe behaves. Researchers once thought they could predict the effects of forces using only Newton's laws (see page 16). However, when astronomers looked far out into space, they found that Newton's rules are only approximately correct and do not work well when forces and speeds are very great. Albert Einstein provided a new way of looking at the Universe, which he called relativity. He explained that both time and space are not constant as Newton believed. Both movement and gravity bend and stretch time and space, although the effects are usually too small to notice on Earth. This view of the Universe is puzzling, but it helps us understand how the Universe may have begun.

Light is reflected straight back by a mirror 9 km (5.5 miles) away

Outward beam passes through gap in teeth

Cog spins at high speed during light's two-way journey and will block the returning beam with its next tooth

Light reflected towards distant mirror

Lamp (source of light)

Observer sees reflected beam

One-way mirror lets reflected beam through

SPEED OF LIGHT
Light seems to move instantly because its movement is far too fast for human eyes to see. In 1849, the French physicist Hippolyte Fizeau tried to measure exactly how fast. He shone a light onto a mirror 9 km (5.5 miles) away, passing the beam through the teeth of a spinning cog on the way. Fizeau increased the speed of the cog but it never got fast enough to stop the outbound light passing through a gap between its teeth. However, at a certain speed, the light coming back from the mirror grew dim. This happened because the rays were blocked on their way back as the next tooth swung in front. Fizeau knew the speed of the cog's teeth, and this told him that the light took only a few millionths of a second to travel the 18 km (11 miles) to the mirror and back – at a speed of about 300,000 km (186,000 miles) per second. This figure was close to modern measurements.

THE MYSTERY OF LIGHT SPEED
Albert Einstein wondered if light coming from a fast-moving object, such as a car, travels faster than a beam of light from a stationary lamp beside the road. In fact, nobody could find any difference in the speed of light from moving sources. The headlight beam from a speeding car reaches a driver heading the other way at exactly the same speed as it leaves the headlights travelling in the opposite direction. For this contradiction to be true, Einstein made the startling suggestion that time and space must change while the speed of light stays the same.

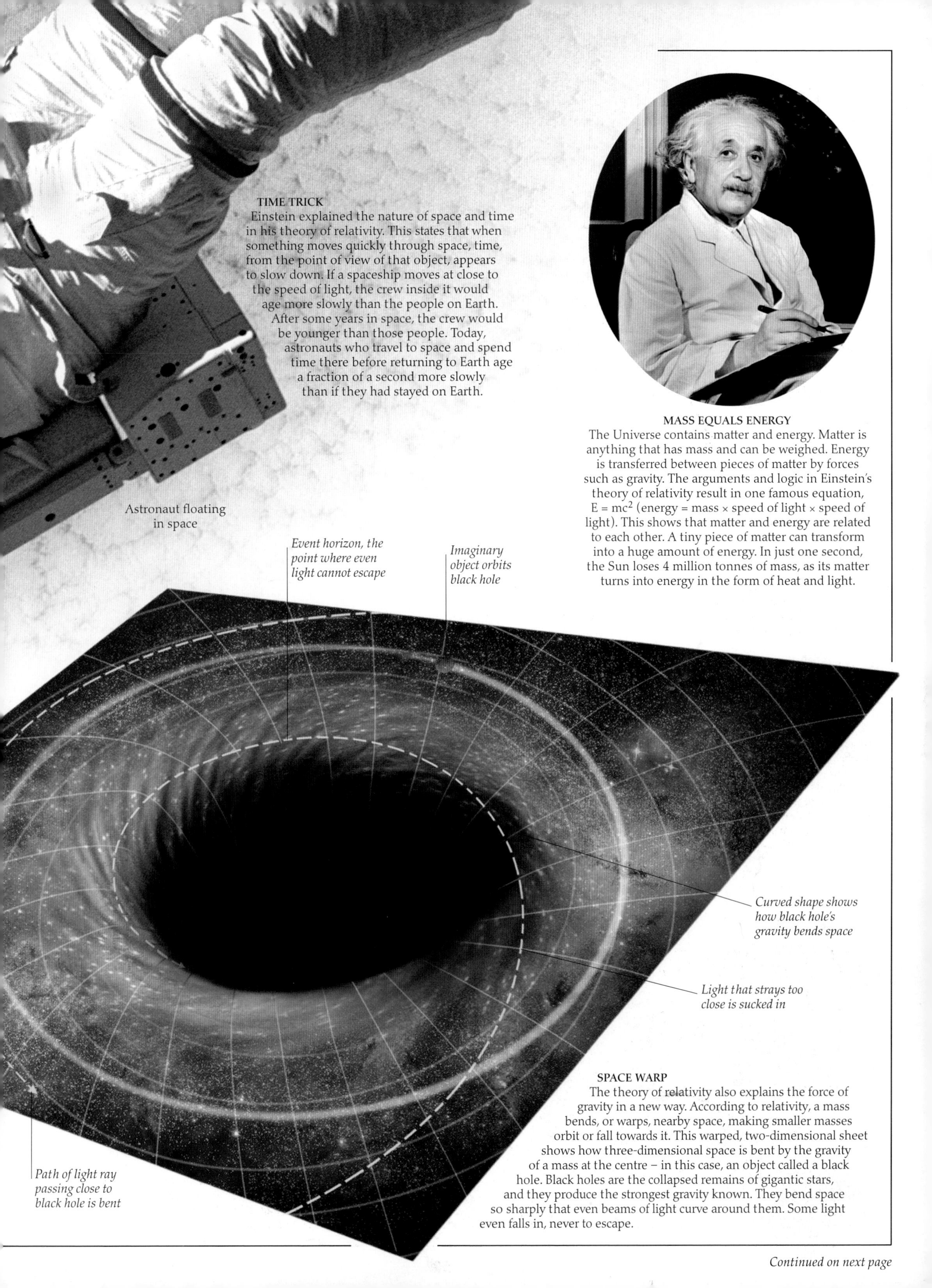

TIME TRICK

Einstein explained the nature of space and time in his theory of relativity. This states that when something moves quickly through space, time, from the point of view of that object, appears to slow down. If a spaceship moves at close to the speed of light, the crew inside it would age more slowly than the people on Earth. After some years in space, the crew would be younger than those people. Today, astronauts who travel to space and spend time there before returning to Earth age a fraction of a second more slowly than if they had stayed on Earth.

Astronaut floating in space

MASS EQUALS ENERGY

The Universe contains matter and energy. Matter is anything that has mass and can be weighed. Energy is transferred between pieces of matter by forces such as gravity. The arguments and logic in Einstein's theory of relativity result in one famous equation, $E = mc^2$ (energy = mass × speed of light × speed of light). This shows that matter and energy are related to each other. A tiny piece of matter can transform into a huge amount of energy. In just one second, the Sun loses 4 million tonnes of mass, as its matter turns into energy in the form of heat and light.

Event horizon, the point where even light cannot escape

Imaginary object orbits black hole

Curved shape shows how black hole's gravity bends space

Light that strays too close is sucked in

Path of light ray passing close to black hole is bent

SPACE WARP

The theory of relativity also explains the force of gravity in a new way. According to relativity, a mass bends, or warps, nearby space, making smaller masses orbit or fall towards it. This warped, two-dimensional sheet shows how three-dimensional space is bent by the gravity of a mass at the centre – in this case, an object called a black hole. Black holes are the collapsed remains of gigantic stars, and they produce the strongest gravity known. They bend space so sharply that even beams of light curve around them. Some light even falls in, never to escape.

Continued on next page

Continued from previous page

VAST UNIVERSE
The Universe is so enormous that it can take years for light to travel from one star to another nearby. The light from the Sun takes 8 minutes to reach Earth. This means that when we look at the Sun, we see it as it looked 8 minutes ago. The farthest objects astronomers have seen so far are about 13 billion light years away. This is the distance a beam of light travels in 13 billion years. Stars that far away look like they did 13 billion years ago – most have probably disappeared now. So looking farther out into space is like seeing back in time as well. The Hubble Space Telescope takes pictures of stars and galaxies (clusters of stars). This picture – called the Ultra Deep Field – was taken by the Hubble telescope and shows a region of space with some of the oldest, most distant galaxies.

RED SHIFT
The Universe is constantly expanding. Scientists discovered this when they noticed that the wavelengths of light coming from very distant stars are redder than expected – a phenomenon called red shift. They explained this by suggesting that the space between stars is stretching. As space stretches, the wavelengths of starlight shining through it are stretched as well, and become longer. The longer the wavelength of light, the redder it appears. In every direction, the most distant stars have the most red-shifted light, showing the Universe is expanding everywhere.

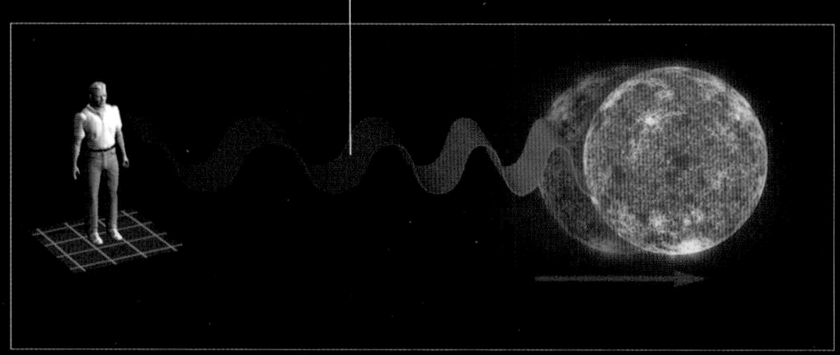
Light from star is stretched in wavelength

Observer

Star moves away as space expands

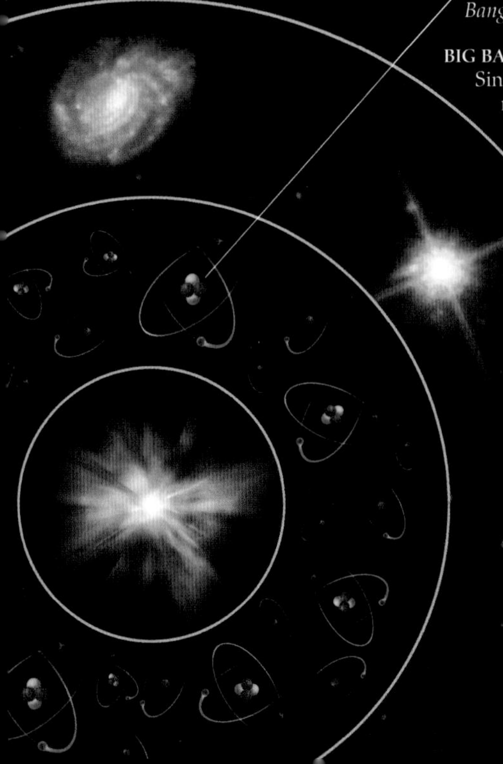
First atoms formed 300,000 years after the Big Bang when the Universe had cooled enough

First stars and galaxies formed as gravity pulled matter together after about 200 million years

BIG BANG
Since the Universe is expanding now, it must have been much smaller in the past. Scientists think that about 13.7 billion years ago, the Universe – including all of its matter and energy – was the size of a grapefruit. This was a split second after what they call the Big Bang – an incredibly violent expansion that released the hot Universe in all directions. Eventually the Universe cooled enough to form atoms, then stars, then planets. This diagram shows the Universe's first 200 million years in three stages.

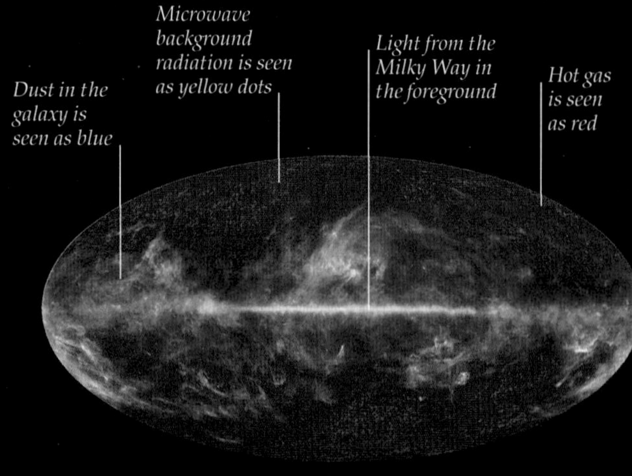
Dust in the galaxy is seen as blue

Microwave background radiation is seen as yellow dots

Light from the Milky Way in the foreground

Hot gas is seen as red

ECHOES FROM THE PAST
The Big Bang produced a lot of radiation, and amazingly we can still detect some of it. In 1964, astronomers picked up faint microwave signals coming from space. They found that they could detect the same radiation wherever they looked. This radiation was termed the Cosmic Microwave Background (CMB). Using data from satellites, astronomers have created a map showing how the CMB is spread out across the sky.

SOLAR SYSTEM

The young Universe was made nearly entirely of the gases hydrogen and helium, and these created the first stars. Stars burn by nuclear fusion, creating not only more helium (see page 27), but more elements, such as oxygen and iron. When a star explodes at the end of its life, it sends this rich mixture of material into space ready to create the next stars. It is this stardust that Earth's solar system is made of. Gravity created a whirling spiral of material, with dense hydrogen lighting up to form our Sun at the centre, while clouds of dust and ice further out merged into a string of planets.

END OF DAYS

Stars age too slowly to be tracked during the average human lifetime. After observing many stars at different points in their life, scientists have gathered enough information to accurately predict the life cycle of a star. Based on this they know how stars will change with age. Solar probes launched by NASA and other space agencies study the Sun and its properties. Scientists know the mass of the Sun and the amount of hydrogen it burns every second. They can not only date how old the Sun is, but also find out how long it will last. In about 5 billion years, the Sun will run out of hydrogen fuel and start burning helium and other elements instead. It will swell into a red giant.

CONSTANT COMPANION

Planets such as Jupiter and Neptune have dozens of moons orbiting them, but Earth has just one, and it is very large. Scientists studying moon rocks have found that the composition of the Moon is similar to Earth. This suggests that the Moon was probably created when another planet-sized object smashed into Earth about 4 billion years ago. The impact flung red-hot molten lumps of Earth's surface into space. Earth's gravity kept these lumps in orbit around it, and the pieces eventually merged and cooled to form the Moon.

DARK MATTER

Much of the Universe is not made of visible matter. Astronomers measuring how galaxies spin found that they rotate too fast – as if they were three times as heavy as they thought. These galaxies must contain something heavy but invisible, which the astronomers call dark matter. Scientists may be able to detect dark matter if, as some suspect, it is made up of particles called neutrinos, which are smaller than an atom. Neutrinos are hard to find and large detectors for finding them, such as this one, are built deep inside mountains or mines to reduce the interference from background radiation, such as CMB. The answer to the dark matter mystery could reveal the fate of the Universe. If dark matter is very abundant, its gravity could reverse the Universe's expansion and cause it to shrink back to the size of a grapefruit many billions of years from now.

Unanswered questions

No one knows if science can answer every question. There are still many things scientists do not yet understand, including how particles inside atoms behave. The scientists can measure the particles moving, and determine their positions, but cannot do both things at the same time. It may turn out that they are not even particles at all! The latest research also suggests that the rules of the Universe might be different in distant parts of space. Gravity and other forces might not be fixed, and may change in ways that we do not yet understand. Many great scientists have searched for a "theory of everything" that could explain these problems, but so far no answer has been found, just more questions.

MACHINE INTELLIGENCE

Computer scientists are building robots, such as this one with Albert Einstein's face, that can learn using artificial intelligence (AI). No one knows if a future AI machine could think like a human. A microchip computer with as much memory and processing power as a human brain would be larger than a skyscraper. Even then, it may not be aware of itself, nor generate true feelings like a human. Could a robot really understand like humans do? A quantum computer, which uses single atoms as switches, might match a human brain for size and speed, but scientists are still learning how to build one.

Wires connect robot to computer brain that learns how to use the correct facial expressions

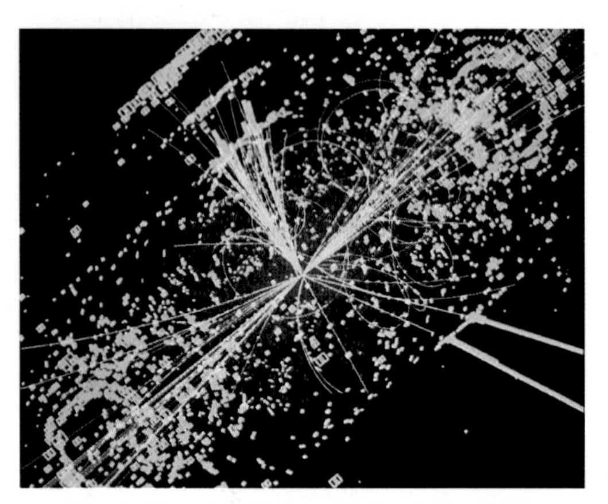

Illustration of particles colliding in the LHC

LITTLE BIG BANG

Stars burn because some of their atom's mass turns into pure energy during nuclear fusion (see page 47). But scientists do not understand the opposite process – how energy can form atoms. They think this is what happened in the Big Bang. Physicists at CERN, an immense underground laboratory in Switzerland, are trying to recreate the conditions of the Big Bang. In one experiment, a machine called the Large Hadron Collider (LHC, see page 7) fired streams of particles at each other at nearly the speed of light, making them collide. Scientists then recorded the collisions as streaked patterns. In future experiments, they hope to observe for the first time how matter forms from pure energy.

Spike indicates a powerful source of radio waves

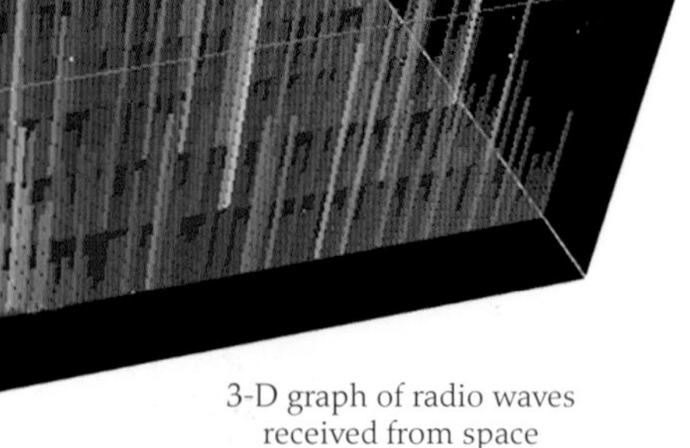

3-D graph of radio waves received from space

MEET THE NEIGHBOURS

Humans sent the first signals into space about 80 years ago. This was a side-effect of our radio signals becoming powerful enough to reach beyond the atmosphere. If there are alien civilizations near distant stars, then they may produce similar signals. Scientists working for a project called SETI – the Search for Extraterrestrial Intelligence – use computers to search through the din of natural radio waves coming from space. They look for signals that contain patterns. Even though radio waves travel at the speed of light, signals from distant parts of space will take a long time to reach Earth. An alien signal – if detected – would probably be hundreds or thousands of years old.

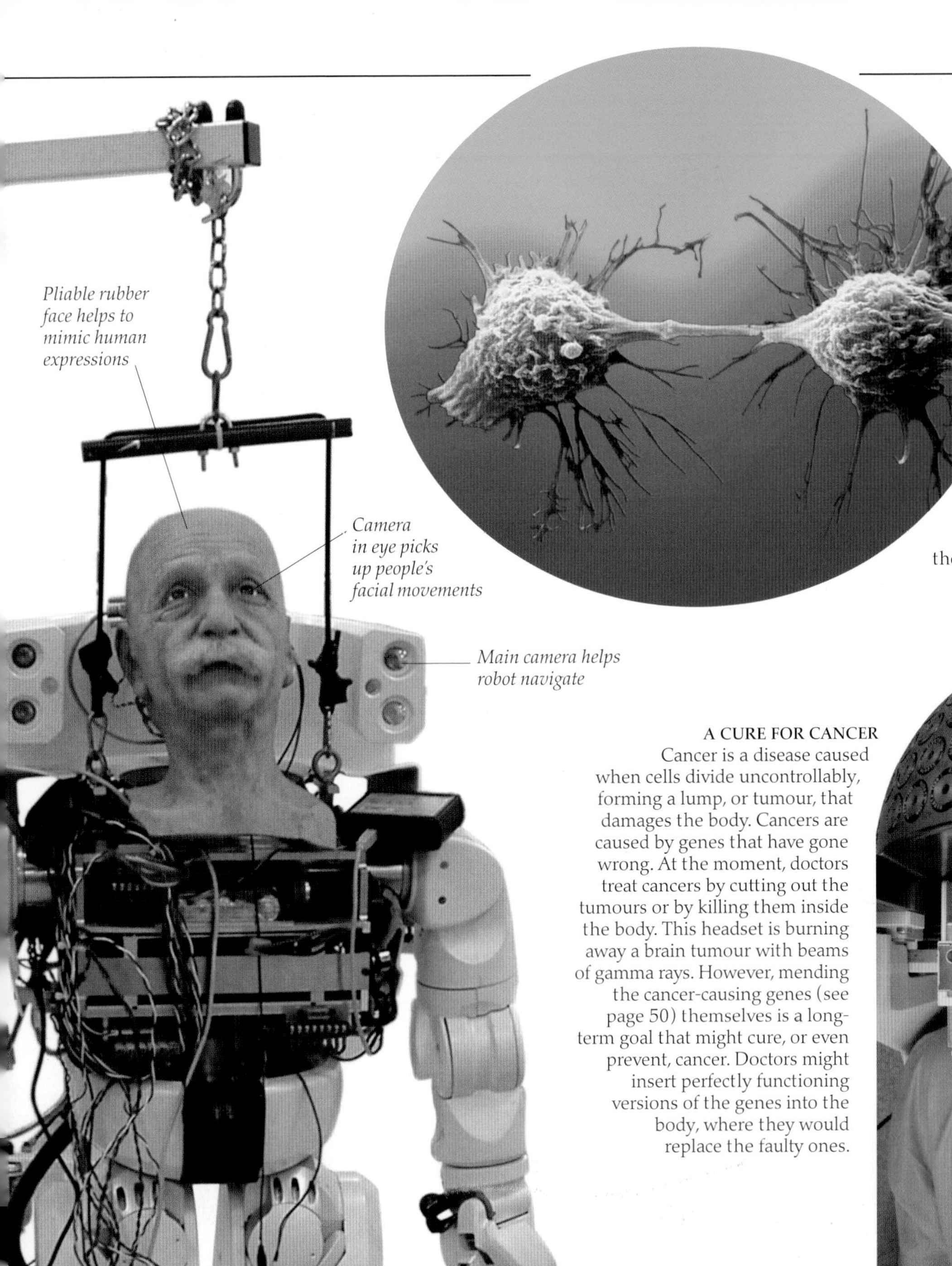

The robot Albert Einstein

IMMORTAL LIFE

New cells form when one cell doubles up its contents and splits in half forming two identical cells. Body cells have in-built clocks that stop them from dividing after they have split a certain number of times. Eventually, the body cannot grow enough new cells to stay healthy and it will die of old age. If scientists could switch off a person's cell clocks, he or she might be able to live forever. Scientists keep living colonies of human cells in laboratories for use in experiments. Some of them, like the HeLa cells shown here, can divide indefinitely. Can scientists find out how to give the immortality of HeLa cells to humans?

A CURE FOR CANCER

Cancer is a disease caused when cells divide uncontrollably, forming a lump, or tumour, that damages the body. Cancers are caused by genes that have gone wrong. At the moment, doctors treat cancers by cutting out the tumours or by killing them inside the body. This headset is burning away a brain tumour with beams of gamma rays. However, mending the cancer-causing genes (see page 50) themselves is a long-term goal that might cure, or even prevent, cancer. Doctors might insert perfectly functioning versions of the genes into the body, where they would replace the faulty ones.

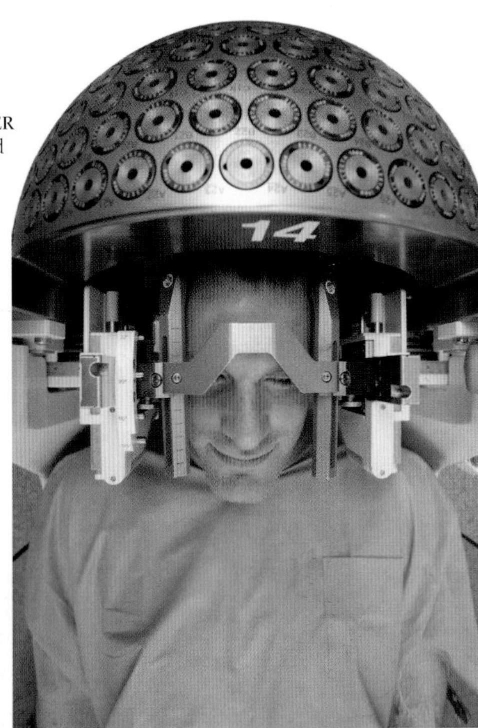

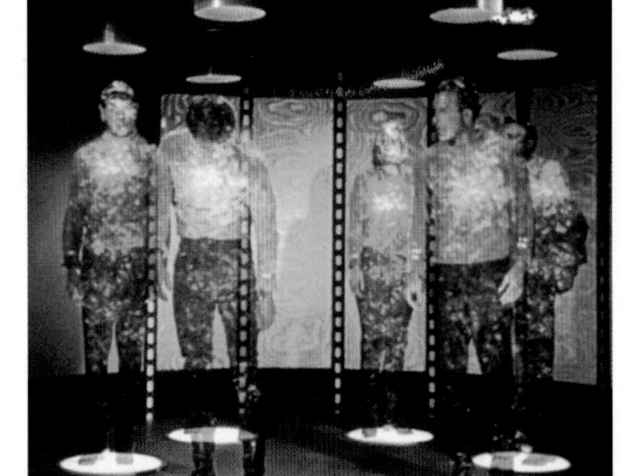

TELEPORTATION

Science fiction can sometimes turn into fact. In the 1960s, the TV series *Star Trek* featured people who instantly travelled or beamed using a machine called a teleporter. Scientists think they could invent it one day by using the strange properties of subatomic particles. Such particles exist in pairs and behave identically even when far apart. A teleporter would break down an object into its many particles. This change would be experienced by each of the particles' pairs in another location. Then the machine would rebuild the object using the second set of particles, making an identical version at a different place.

Periodic table

The periodic table contains all the elements yet discovered – 118 so far, although only 92 of them are found naturally on Earth. The table lists the elements in order of their atomic number – the number of protons in the nuclei of their atoms (see pages 24–25). It also arranges the elements in groups (columns) and periods (rows). The periodic table is divided into sections that tell chemists what each element looks like and how it reacts with other elements.

Atomic number is the number of protons in the nucleus

Symbol is shorthand for the element's name

26
Fe
Iron
56

Mass number is the sum of protons and neutrons in the nucleus

SYMBOL
Every element has a unique chemical symbol, which can either be key letters of the element's name, or letters from older terms for the element. For example, iron's symbol, Fe, comes from *ferrum*, the Latin word for the metal, while lead's symbol is Pb, as its original Latin name is *plumbum*. Each element is also described by its atomic number and a second, larger mass number.

Group One contains the most reactive metals

TABLE OF ELEMENTS
The periodic table has eight main groups. Group One elements have just one electron in their outermost atomic shell, while Group Eight elements have eight electrons. The members of a group react in a similar way because they all have the same number of outer electrons. The periods (rows) show how heavy an atom is – and how likely it is to react with other elements. Heavy metal atoms react more often than the light ones, while, in general, light non-metals react more often than the heavier ones.

1 H Hydrogen 1							
3 Li Lithium 7	4 Be Beryllium 9						
11 Na Sodium 23	12 Mg Magnesium 24						
19 K Potassium 39	20 Ca Calcium 40	21 Sc Scandium 45	22 Ti Titanium 48	23 V Vanadium 51	24 Cr Chromium 52	25 Mn Manganese 55	26 Fe Iron 56
37 Rb Rubidium 85	38 Sr Strontium 88	39 Y Yttrium 89	40 Zr Zirconium 90	41 Nb Niobium 93	42 Mo Molybdenum 98	43 Tc Technetium 97	44 Ru Ruthenium 102
55 Cs Caesium 133	56 Ba Barium 138	57–71	72 Hf Hafnium 180	73 Ta Tantalum 181	74 W Tungsten 184	75 Re Rhenium 187	76 Os Osmium 192
87 Fr Francium 223	88 Ra Radium 226	89–103	104 Rf Rutherfordium 261	105 Db Dubnium 262	106 Sg Seaborgium 268	107 Bh Bohrium 264	108 Hs Hassium 267

Group Two metals are commonly found in rocks.

57 La Lanthanum 139	58 Ce Cerium 140	59 Pr Praseodymium 141	60 Nd Neodymium 142	61 Pm Promethium 145
89 Ac Actinium 227	90 Th Thorium 232	91 Pa Protactinium 231	92 U Uranium 238	93 Np Neptunium 237

The Lanthanide and Actinide series have a special position on the table

REACTIVE METALS
By the 19th century, chemists saw that the elements could be grouped according to their properties. Dmitri Mendeleev set out these groups in his periodic table in 1869. Chemists now know that the members of each group have atoms with the same number of outer electrons. Group One atoms have just one and that makes them very likely to react. Most of Group One is made up of the alkali metals. Hydrogen is not a metal and is placed with this group for convenience. Alkali metals form strong alkali compounds, and many salts, such as common salt. Sodium and potassium are common alkali metals and are important for keeping the body's muscles and nerves healthy.

Potassium burns with a distinctive lilac-coloured flame

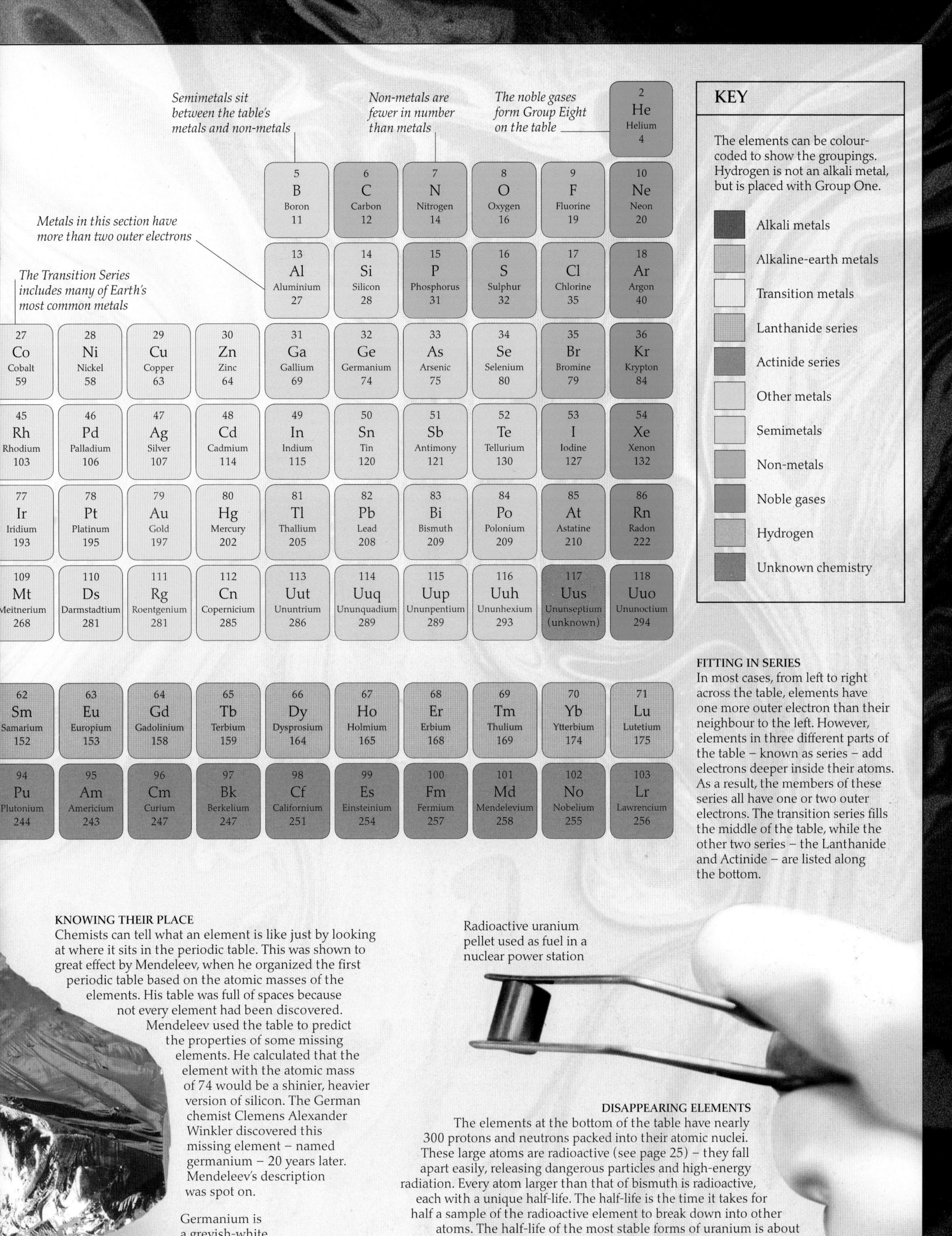

Semimetals sit between the table's metals and non-metals

Non-metals are fewer in number than metals

The noble gases form Group Eight on the table

Metals in this section have more than two outer electrons

The Transition Series includes many of Earth's most common metals

									2 He Helium 4
				5 B Boron 11	6 C Carbon 12	7 N Nitrogen 14	8 O Oxygen 16	9 F Fluorine 19	10 Ne Neon 20
				13 Al Aluminium 27	14 Si Silicon 28	15 P Phosphorus 31	16 S Sulphur 32	17 Cl Chlorine 35	18 Ar Argon 40
27 Co Cobalt 59	28 Ni Nickel 58	29 Cu Copper 63	30 Zn Zinc 64	31 Ga Gallium 69	32 Ge Germanium 74	33 As Arsenic 75	34 Se Selenium 80	35 Br Bromine 79	36 Kr Krypton 84
45 Rh Rhodium 103	46 Pd Palladium 106	47 Ag Silver 107	48 Cd Cadmium 114	49 In Indium 115	50 Sn Tin 120	51 Sb Antimony 121	52 Te Tellurium 130	53 I Iodine 127	54 Xe Xenon 132
77 Ir Iridium 193	78 Pt Platinum 195	79 Au Gold 197	80 Hg Mercury 202	81 Tl Thallium 205	82 Pb Lead 208	83 Bi Bismuth 209	84 Po Polonium 209	85 At Astatine 210	86 Rn Radon 222
109 Mt Meitnerium 268	110 Ds Darmstadtium 281	111 Rg Roentgenium 281	112 Cn Copernicium 285	113 Uut Ununtrium 286	114 Uuq Ununquadium 289	115 Uup Ununpentium 289	116 Uuh Ununhexium 293	117 Uus Ununseptium (unknown)	118 Uuo Ununoctium 294
62 Sm Samarium 152	63 Eu Europium 153	64 Gd Gadolinium 158	65 Tb Terbium 159	66 Dy Dysprosium 164	67 Ho Holmium 165	68 Er Erbium 168	69 Tm Thulium 169	70 Yb Ytterbium 174	71 Lu Lutetium 175
94 Pu Plutonium 244	95 Am Americium 243	96 Cm Curium 247	97 Bk Berkelium 247	98 Cf Californium 251	99 Es Einsteinium 254	100 Fm Fermium 257	101 Md Mendelevium 258	102 No Nobelium 255	103 Lr Lawrencium 256

KEY

The elements can be colour-coded to show the groupings. Hydrogen is not an alkali metal, but is placed with Group One.

- Alkali metals
- Alkaline-earth metals
- Transition metals
- Lanthanide series
- Actinide series
- Other metals
- Semimetals
- Non-metals
- Noble gases
- Hydrogen
- Unknown chemistry

FITTING IN SERIES

In most cases, from left to right across the table, elements have one more outer electron than their neighbour to the left. However, elements in three different parts of the table – known as series – add electrons deeper inside their atoms. As a result, the members of these series all have one or two outer electrons. The transition series fills the middle of the table, while the other two series – the Lanthanide and Actinide – are listed along the bottom.

KNOWING THEIR PLACE

Chemists can tell what an element is like just by looking at where it sits in the periodic table. This was shown to great effect by Mendeleev, when he organized the first periodic table based on the atomic masses of the elements. His table was full of spaces because not every element had been discovered. Mendeleev used the table to predict the properties of some missing elements. He calculated that the element with the atomic mass of 74 would be a shinier, heavier version of silicon. The German chemist Clemens Alexander Winkler discovered this missing element – named germanium – 20 years later. Mendeleev's description was spot on.

Germanium is a greyish-white semimetal

Radioactive uranium pellet used as fuel in a nuclear power station

DISAPPEARING ELEMENTS

The elements at the bottom of the table have nearly 300 protons and neutrons packed into their atomic nuclei. These large atoms are radioactive (see page 25) – they fall apart easily, releasing dangerous particles and high-energy radiation. Every atom larger than that of bismuth is radioactive, each with a unique half-life. The half-life is the time it takes for half a sample of the radioactive element to break down into other atoms. The half-life of the most stable forms of uranium is about 4.5 billion years, while the half-life of francium is just 22 minutes.

Measuring things

SCIENCE WOULD BE IMPOSSIBLE without any measurements. It is crucial that all scientists measure things in exactly the same way, so they can compare what they have found. Different measurements require different units, many of which are named after famous scientists. For example, force is measured in newtons (N), in honour of Isaac Newton. These graphics show a few measurement units and the things they measure.

UNITS OF MEASUREMENT
All measurements are built on seven base units, known as the SI units. The units measure the fundamental characteristics of the Universe. Every other unit of measurement is a combination of two or more of these base units. For example, the newton is a combination of metre, second, and kilogram.

Unit	Symbol	Definition
metre	m	The unit of distance. Metres are defined using light. Light travels a distance of one metre in 0.3 billionths of a second.
kilogram	kg	The unit of mass. A kilogram is the mass of a litre of water. There are 1,000 litres in a cubic metre.
second	s	The unit of time. There are 86,400 seconds in a day, and scientists measure seconds by counting the vibrations of certain atoms.
kelvin	K	The unit of temperature. 0 K is the lowest temperature possible. It is also known as absolute zero.
candela	cd	The unit of light intensity. A burning candle emits about one candela of light.
ampere	A	The unit of electricity. An ampere, or amp, of electricity moves about 6 million trillion electrons every second.
mole	mol	The unit of quantity. One mole contains 6,000 billion trillion atoms or molecules.

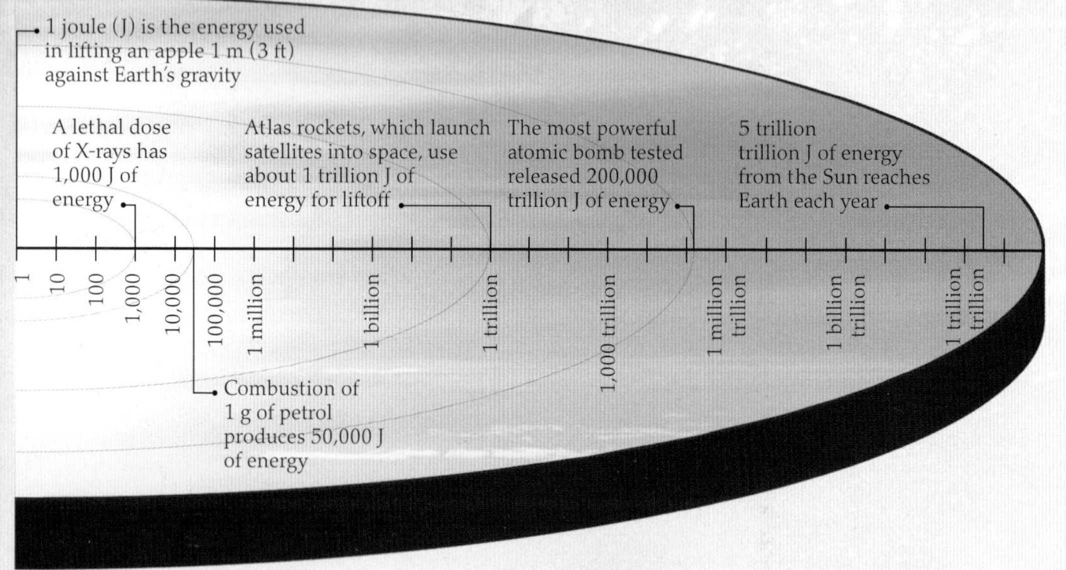

OUTPUT OF ENERGY
Energy is measured in joules, named after James Prescott Joule, a scientist who calculated how heat energy translates into motion. One joule (J) is the energy needed to lift 1 kg (2.21 lb) by 10 cm (3.9 in) in 1 second. Energy is difficult to understand. Everything possesses energy but it cannot be seen. Energy is not created or destroyed, but just transferred from one object – an atom or a ball, for example – to another. The most familiar form of energy is heat, but it can also be observed as light, motion, and electricity.

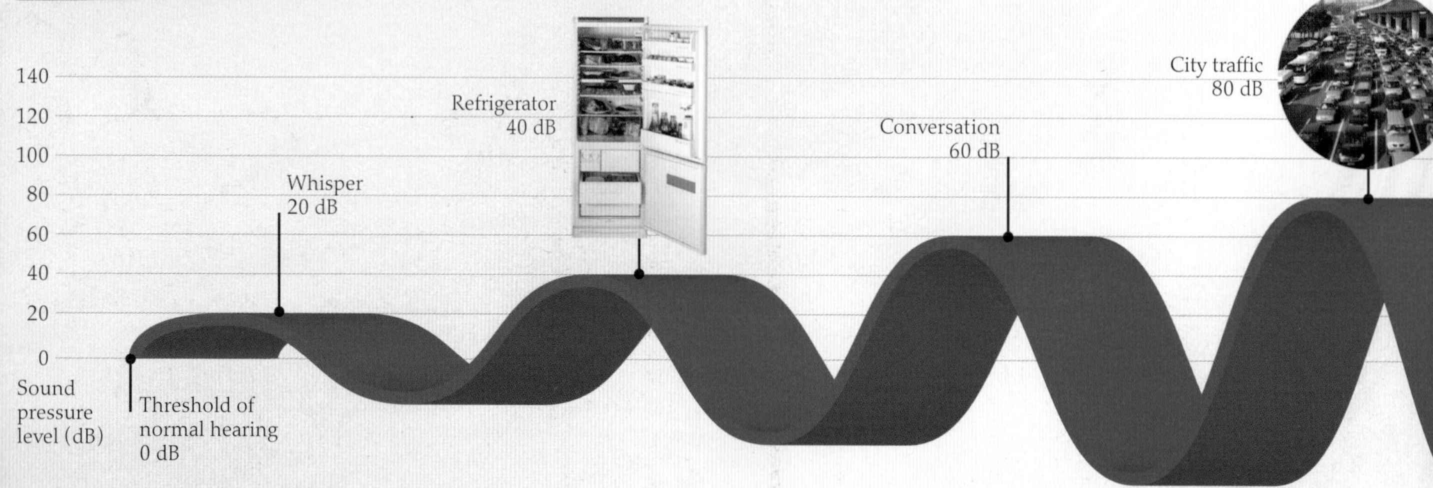

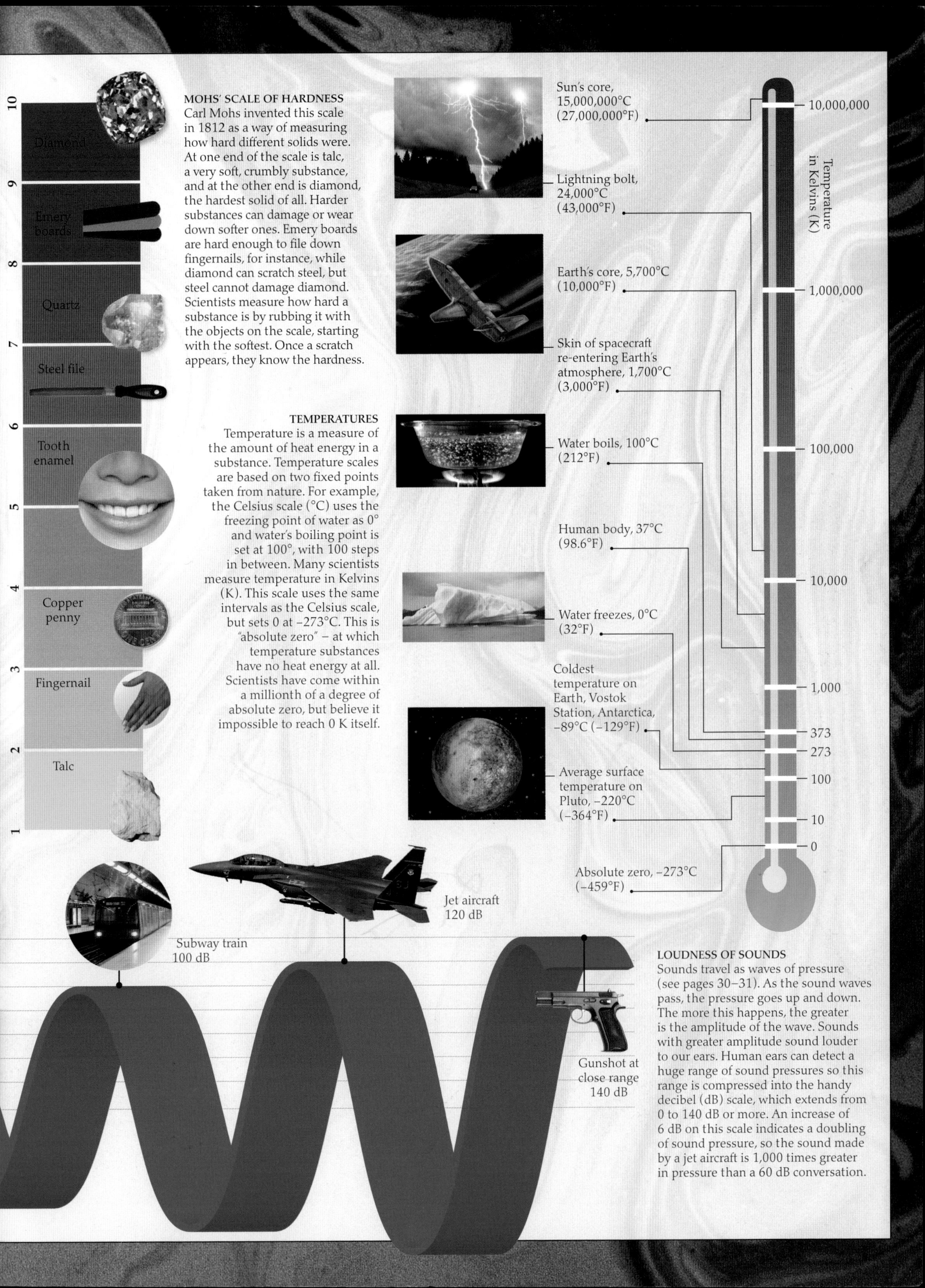

MOHS' SCALE OF HARDNESS

Carl Mohs invented this scale in 1812 as a way of measuring how hard different solids were. At one end of the scale is talc, a very soft, crumbly substance, and at the other end is diamond, the hardest solid of all. Harder substances can damage or wear down softer ones. Emery boards are hard enough to file down fingernails, for instance, while diamond can scratch steel, but steel cannot damage diamond. Scientists measure how hard a substance is by rubbing it with the objects on the scale, starting with the softest. Once a scratch appears, they know the hardness.

TEMPERATURES

Temperature is a measure of the amount of heat energy in a substance. Temperature scales are based on two fixed points taken from nature. For example, the Celsius scale (°C) uses the freezing point of water as 0° and water's boiling point is set at 100°, with 100 steps in between. Many scientists measure temperature in Kelvins (K). This scale uses the same intervals as the Celsius scale, but sets 0 at –273°C. This is "absolute zero" – at which temperature substances have no heat energy at all. Scientists have come within a millionth of a degree of absolute zero, but believe it impossible to reach 0 K itself.

LOUDNESS OF SOUNDS

Sounds travel as waves of pressure (see pages 30–31). As the sound waves pass, the pressure goes up and down. The more this happens, the greater is the amplitude of the wave. Sounds with greater amplitude sound louder to our ears. Human ears can detect a huge range of sound pressures so this range is compressed into the handy decibel (dB) scale, which extends from 0 to 140 dB or more. An increase of 6 dB on this scale indicates a doubling of sound pressure, so the sound made by a jet aircraft is 1,000 times greater in pressure than a 60 dB conversation.

Fields of science

SCIENTISTS CAN STUDY almost anything – from distant stars to tiny bacteria. No one can be highly knowledgeable about all these things, and so most scientists specialize in just one field of science. The fields fall roughly into six groups – biological sciences, earth sciences, physical sciences, social sciences, astronomy, and mathematics. However, there is a lot of overlap between fields, and different specialists often work together to study complex problems.

Ecology
Understanding how different life forms rely on each other and on their environment for survival

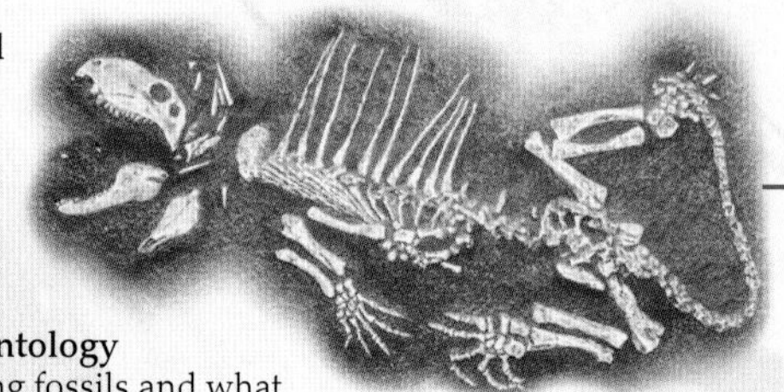

A fossilized reptile skeleton from 250 million years ago

Palaeontology
Studying fossils and what they tell us about living organisms in the past, including how life began and how it has evolved

Environmental science
Using a range of information from biology, chemistry, and geology to understand how all the different parts of the environment work together

Computers process data as 1s and 0s

Computer science
A branch of maths investigating how people can program computers and how computers run their programs, including how they process and output data

Anthropology
Investigating the variety in the human species, including how humans and their different cultures have changed

Mathematics
A tool scientists use, but also a field of study that looks at relationships between numbers

Social sciences
Using scientific methods to explain how humans behave on their own, in groups, or in society

Biological sciences
Studying things that are alive, or were once alive – these sciences are also known as life sciences

Psychology
Studying the human mind, or psyche, and why people behave the way they do, often to help people cope with mental stress and illnesses

Sigmund Freud (1856–1939), pioneer of psychology

Archaeology
Investigating traces of human life in the ground, including bones, tools, and ruins, to figure out how people lived in the past

Human geography
Researching how humans use the landscape, including the patterns of their settlement, transport, and activities

Economics
Examining how people, companies, and governments decide which goods and services to produce or consume

Pathology
Researching diseases, how they attack the body, and how they can be stopped

Anatomy
Figuring out the structure of a body and how it works

Botany, or plant science
Studying plants, including their variety, structure, chemical processes, and their life in the wild

Epidemiology
Tracking the spread of diseases, including fast-spreading epidemics, from person to person, through a community, and around the world

Medicine
Finding cures and treatments for illnesses and injuries, generally using chemical medicines, and physical treatments, such as surgery

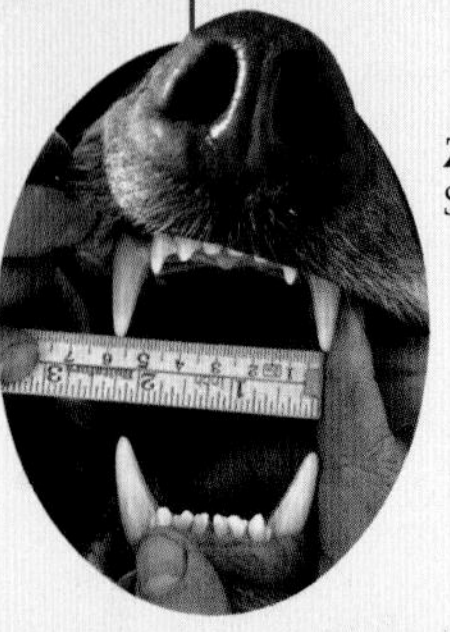

Zoologist measures a bear's teeth

Zoology
Studying animals, including how they survive and reproduce in the wild

Microbiology
Examining the smallest living things, using microscopes to see their details

Geology
Studying how Earth's rocky surface layer has formed and changed over the life of the planet

Meteorology
Understanding Earth's different climates and predicting changes in the weather

Robotics
A field of engineering aiming to build machines that can move, sense their surroundings, and think for themselves

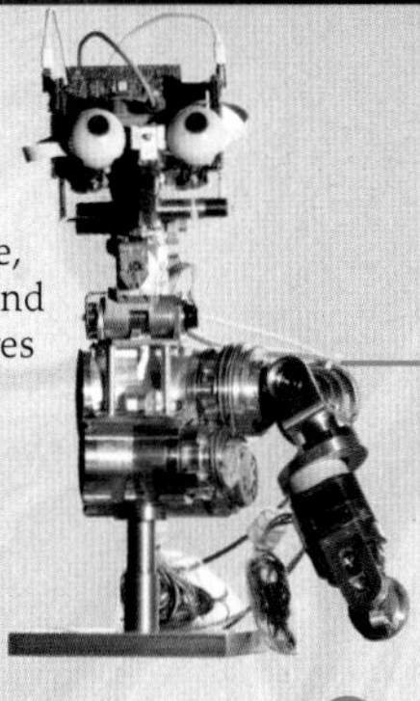

iCub, a robot that can learn

Engineering
Using scientific understanding to build useful machines

Acoustics
Investigating the behaviour of sound waves

Mechanics
Studying how forces make objects move

Physical geography
Studying the formation and structure of Earth's surface features

Oceanography
Learning about all aspects of oceans and seas, including tides, currents, the chemistry of the water, and marine life

Researchers studying marine life

Nuclear physics
Looking inside an atom's nucleus, particularly during radioactive decay, nuclear fission, and nuclear fusion

Electronics
Examining the flow of electric currents through substances, and how to use them to run microchips, computers, and other devices

Chemistry
Studying how atoms connect to produce new substances in reactions – and how those substances behave

Optics
Investigating the behaviour of light – how it bends, reflects, and scatters

Earth sciences
Studying the non-living parts of Earth, such as rocks, volcanoes, and the atmosphere

Astronomy
Studying objects in the Universe beyond Earth, such as stars, planets, and galaxies

Physical sciences
Studying non-living things, such as atoms, energy, and radiation

Cosmology
Learning about the Universe as a whole, including how it formed, its shape and size, and how it may end

Astrophysics
Studying the physics of stars, black holes, galaxies, and other massive objects in space

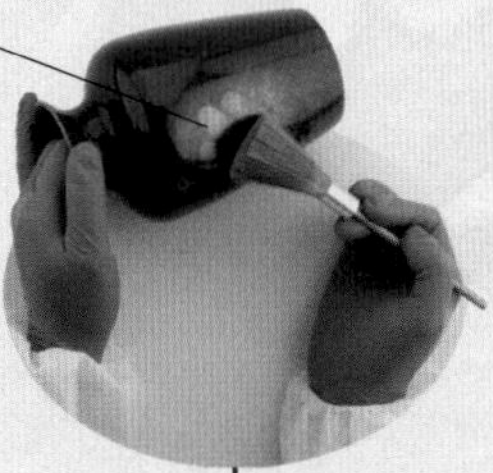

Fingerprints show up in fine dust

Forensic scientist dusting for fingerprints

Forensic science
Using a range of scientific knowledge to investigate evidence from crime scenes

Geophysics
Using knowledge of physics to investigate the structure of planets

Biochemistry
Studying the chemical reactions that drive life processes

Genetics
Studying how characteristics controlled by genes are passed on from one generation to another and how they vary in living things

Red blood cell

Cell biology and Molecular biology
Understanding how cells inside living organisms work – both as a whole and at the level of their DNA, proteins, and other complex molecules

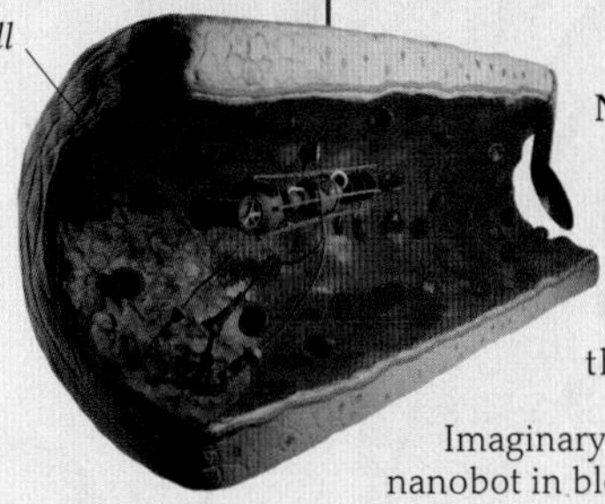

Nanotechnology
Using engineering, molecular biology, and other knowledge to invent machines, such as nanobots, no bigger than a human cell

Imaginary future nanobot in blood vessel

Immunology
Examining the immune system of humans and other animals and how it defends the body against attack from diseases

Biophysics
Using chemistry and physics to study biological systems, such as animal senses and the way animals move

Glossary

ACID
A reactive chemical that contains a hydrogen ion. This breaks off from the rest of the molecule easily and attacks other chemicals.

ALCHEMY
A way of studying nature, prior to the modern scientific method. Alchemists believed that magic was an important part of natural processes.

ALKALI
A substance that reacts most strongly with acids. An alkali contains a hydroxide ion that combines with an acid's hydrogen ion to make water.

ALLOY
A mixture of different metals. Many alloys are stronger and longer-lasting than pure metals. Common alloys include brass and steel.

AMINO ACID
A simple chemical that links into long chains to make complex protein molecules. Twenty-one different amino acids occur naturally.

APPARATUS
The equipment used by scientists in a laboratory. Simple apparatus includes gas burners, glass beakers (cups), and test tubes.

ATOM
The smallest unit of an element. Every element has its own type of atom, with a unique size, weight, and structure.

ATOMIC BOMB
A bomb that explodes due to a nuclear reaction inside – either fission (big atoms splitting) or fusion (small atoms joining together).

BACTERIUM
One of many tiny single-celled life forms (plural bacteria) that live worldwide, even under extreme heat or pressure. Bacteria have existed on Earth for at least 3.5 billion years.

Camouflage among leaves

BAROMETER
A device for measuring the pressure of the air pushing at Earth's surface.

BIG BANG
The event that most scientists think started the Universe nearly 14 billion years ago. It created all the matter and energy in the Universe.

BLACK HOLE
A massive object in space with enough gravity to pull in all things around it, even light.

CAMOUFLAGE
A method of survival used by some organisms, in which they change their appearance to blend into their environment and avoid detection.

CELL
The building blocks of a living body. Cells are self-contained units, but work together.

CHEMICAL REACTION
The process in which atoms from two or more chemicals arrange themselves into new combinations, making a new set of chemicals.

CHLOROPHYLL
The green-coloured chemical inside many plant cells that absorbs energy from sunlight so it can be used to make sugar food for the plant.

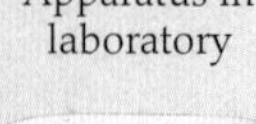

Apparatus in laboratory

CHROMOSOME
The storage unit for a cell's DNA. It lives in the cell's nucleus.

CIRCUIT
A system that allows an electric current to flow through it. All electrical equipment, from light bulbs to computers, are part of circuits.

COMBUSTION
Also known as burning, a chemical reaction involving oxygen and producing light and heat.

COMPOUND
A substance that is formed from the atoms of two or more elements. Water is a compound of hydrogen and oxygen.

DIFFUSION
A gradual process of mixing in fluids (gases and liquids). During diffusion, fluid atoms or molecules move from places where they are densely clustered to places where they are scarce.

DNA
The complex spiral molecule that contains genes (instructions) for making an organism. It is also called deoxyribonucleic acid. Every living thing grows according to instructions in its DNA.

ECHOLOCATION
A method of sensing the surroundings by echoing sounds off surrounding surfaces. Bats use it where it is too dark to see.

ELECTRICITY
The movement of electrons or other charged particles. Electricity flows between areas that have opposite charges.

ELECTROMAGNETIC SPECTRUM
The range of electromagnetic radiation released by atoms. The spectrum includes radio waves, heat, visible light, ultraviolet light, and X-rays.

ELECTROMAGNETISM
The study of electricity and magnetism. These two things are linked. Magnets can create electric currents, and electric currents can make magnets.

ELECTRON
A tiny, negatively charged particle that is found in every atom. Electrons are shared or swapped by atoms during chemical reactions.

ELEMENT
A pure substance that cannot be broken down into simpler ingredients. There are about 90 natural elements, including carbon and oxygen.

ENZYME
A protein – chemical made up of amino acids – that controls one of the many chemical reactions inside a living body.

EVOLUTION
The gradual change in living organisms that occurs over many generations. It is controlled mainly by the process of natural selection, by which the genes that help a life form to survive are passed on to its offspring. When populations become separated, they may start to evolve independently, and new species form.

EXTINCTION
When a species has died out completely, it is extinct. All members are dead and so there will never be new individuals.

FOOD CHAIN
A way of showing how food and energy passes from one life form to another. Food chains start with plants, which get their energy from sunlight. Some animals eat plants and, in turn, get eaten by meat-eating animals.

Jelly dessert, a gel

FREQUENCY
A measure of how often something happens. Waves have a frequency, which is how many times the wave rises and falls in a second.

GEL
A type of mixture made of a liquid mixed with a solid. The solid gives the gel a fixed shape, but the liquid in it makes it wobbly.

GENE
An instruction in code that is passed from parents to their child. The code of genes is written in DNA, and hundreds of genes work together to build a new body.

GENOME
A complete set of the genes of a life form. Not every member of a species has identical copies of each gene. Instead, individuals have one of several slightly different versions.

GRAVITY
The force that pulls objects together. Massive objects produce a stronger pull of gravity than less heavy ones, so light things fall towards heavier ones.

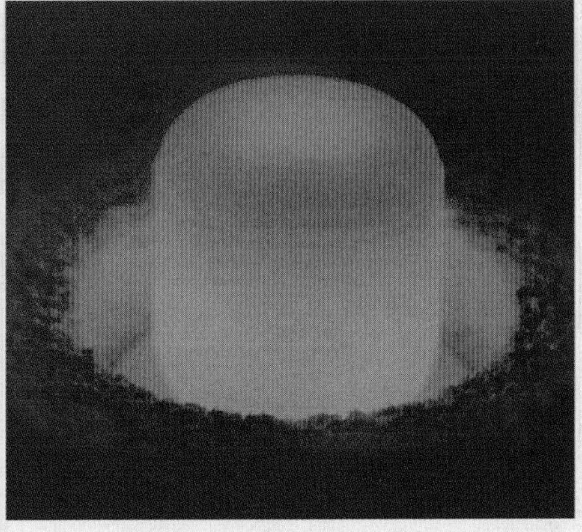

Plutonium glows because of radioactivity

HYDROCARBON
A chemical made up of carbon and hydrogen atoms. Petroleum oil is the main source of common hydrocarbons. Hydrocarbon molecules can have complex structures in the form of chains and rings.

INDUCTION
The process humans use to generate electricity. When magnets move, they create – or induce – a current in any metals nearby.

ION
An atom, or group of connected atoms, which has an electric charge. The charge is created when the atoms give away or gain electrons.

LASER
An artificial light source that emits light of a single, constant wavelength. Unlike those of natural light, the light waves in a laser are lined up with each other.

LENS
A curved, see-through object that bends parallel rays of light so that they seem to pass through a single point.

MAGNET
An object that is normally made from iron or nickel or an alloy, and produces a force field around it that pulls on other pieces of iron, nickel, and a few other metals.

MOLECULE
A group of two or more atoms that are joined together. A molecule is the smallest possible amount of a compound. Breaking it up into smaller units changes the compound.

NEUTRON
A particle found in the central nucleus of almost all atoms. Neutrons are neutral, which means that they do not have an electric charge.

PHOTOSYNTHESIS
The process used by plants to turn the energy in sunlight into sugar food. Photosynthesis is the source of nearly all the sugar in the natural world.

PLASMA
The fourth state of matter, which forms when gases get hot enough that a significant number of molecules have broken into ions and free electrons.

PLASTIC
An artificial substance that retains the shape it is bent into. Most plastics are made from hydrocarbons, and are used to make many products including bottles and parts of spaceships.

POLYMER
A long molecule made from a chain of smaller units. Plastics are polymers, as are DNA, proteins, and the starches in bread and pasta.

PROTON
A positively charged particle found in the nucleus of every atom. Each element has a unique number of protons in its atoms.

RADIATION (ELECTROMAGNETIC)
Waves of energy that are released from atoms – and absorbed by other atoms. Light and radio waves are common forms of radiation.

RADIOACTIVITY
A process that occurs when an atom is unstable, because the protons and neutrons in the nucleus do not stick together. A radioactive atom spontaneously loses small particles from its nucleus – in the form of dangerous radiation – as it becomes more stable.

SEISMIC WAVES
The large-scale vibrations or waves inside Earth's crust that travel through rocks at high speeds. Seismic waves are created when the rocks crack under pressure. When seismic waves reach the surface, they cause earthquakes.

SEMICONDUCTOR
A substance that carries, or conducts, an electric current sometimes, and blocks it at other times. Silicon is a common semiconductor and is used to make the millions of miniature switches in the microchips that run computers.

SPECIES
A group of animals that look very similar and live in the same way. Members of a species must also be able to breed with each other to produce offspring, but not with members of other species.

SUBLIMATION
The process by which a few substances change from a solid state straight into a gaseous state without first melting into a liquid state.

TEMPERATURE
A measure of how much heat energy a substance holds within. Scientists measure temperature using different scales such as the Kelvin scale or the Celsius scale.

Turbine in a power station

TURBINE
A fan-like machine that spins around very fast when a gas or liquid flows through it. Turbines are used in power stations.

ULTRAVIOLET LIGHT
Invisible light that comes from the Sun and causes tanning and sun burn of skin. It is a type of electromagnetic radiation.

WAVELENGTH
A measure of how far a wave moves as it rises and falls to make a single wave shape.

X-RAY
A type of very powerful electromagnetic radiation. X-rays contain a huge amount of energy – enough to shine right through the human body. X-ray beams are used to take pictures of hard objects, such as bones, inside the body.

Index

Acknowledgements

Dorling Kindersley would like to thank: Monica Byles for proofreading and Helen Peters for the index.

The publisher would like to thank the following for their kind permission to reproduce their photographs:

(Key: a-above; b/g-background; b-below/bottom; bl-below left; br-below right; c-centre; cl-centre left; cr-centre right; l-left; r-right; t-top, tl-top left; tr-top right; crb-centre right below; cra-centre right above.)

Alamy Images: Ancient Art & Architecture Collection Ltd 11tc; Phil Degginger 2c, 11tl, 20cl, 64br; JLImages/Cloud Gate, 2004. Anish Kapoor. Stainless steel 33 ft x 66 ft x 42 ft. Millennium Park, Chicago. Courtesy of the City of Chicago and Gladstone Gallery 28b; JR Stock 42br; Mesopotamian / The Art Gallery Collection 6cl; sciencephotos 31t, 40c; Universal Images Group Limited 56cr, 57br; Viennaslide 20c; John Warburton-Lee Photography 49cr; Sue Wilson 55bl; Wiskerke 58bl; **The Bridgeman Art Library:** French School, (20th century) / Private Collection / Archives Charmet 16tl; © **CERN:** 62cl; Maximilien Brice 6-7; **Corbis:** 46-47; Bettmann 12br, 14tr, 46tl, 67ca (space craft), 68cl; G Bowater 71cr; John Carnemolla/Australian Picture Library 18tl; Ralph A. Clevenger 33cb; Richard Cummins 21cr (neon signs); Dr. Richard Kessel & Dr. Gene Shih/Visuals Unlimited 63t; Bruno Ehrs 26cla; Stephen Frink 69cla; Ole Graf 51cr; Justin Guariglia 25tr; Jason Hawkes / Terra 15br; Jakob Helbig / Cultura 27br; Hulton-Deutsch Collection 26crb; Jeff Daly, Inc. / Visuals Unlimited 33br; Koji Kitagawa/ amanaimages 30cl; Rolf Kosecki 29c; Beau Lark 67cl (Smile); Liu Liqun 42cl; Alan Marsh 38ca (liquid hand wash); Rob Matheson 40-41; Joe McDonald 4tr, 31c; Mediscan / Encyclopedia 7br; Charles Melton 53ftr; Micro Discovery 54bl; NASA 47tr; Richard T. Nowitz 33tl; Micha Pawlitzki 66c; PBNJ Productions 64-65, 66-67, 68-69, 70-71 (border); Louie Psihoyos/Science Faction 45cr; Michael Rosenfeld 20bc; Rosner, Arnie/Index Stock 22t; Charles E. Rotkin 44-45; Kevin Schafer 68tr; Denis Scott 67cb (Pluto); Gerhilde Skoberne 28t; Paul Souders 66r; STScI/NASA 61t; The Gallery Collection 14bc; Bruce Benedict/Transtock 44bl (Car); Visuals Unlimited 20tl; William Whitehurst 44tl; Staffan Widstrand 68br; Tim Wright 25crb; **Dorling Kindersley:** Chad Ehlers © Alamy 42bl; The Science Museum, London 4br, 15bl, 30b, 40cl; National Maritime Museum, London 2tr, 16cl, 17c; **ESA:** LFI and HFI Consortia 60br; **Getty Images:** 3D4Medical.com 41br; 3DClinic 57tr; AFP 59tr; Steve Allen 67tr (lightning); Colin Anderson / Brand X Pictures 38-39c (main image); Art Montes De Oca 49br; Jeffrey Coolidge 2cl, 41t; Peter Dazeley 67br; De Agostini Picture Library 67cl (Quartz), 67clb (Talc); Digital Vision 24br, 67cl (Steel file); Wally Eberhart 51tr; Don Farrall 29cla, 29tl; Stephen Frink 55br; Jonathan S Blair/National Geographic 23tr; Tim Graham 32bl; Jorg Greuel 68tl; Huntstock 55tr; Liu Jin / AFP 8-9; Hannah Johnston 10-11; Jupiterimages / Comstock Images 7bl; Kurtwilson 71t; Marwan Naamani / AFP 9tr; Photodisc 38br; Photosindia 41c; Popperfoto 52tl; Science Photo Library/Pasieka 3tl, 51tl; SSPL 15cr; Brian Stablyk 44bl (truck); Stocktrek Images 67bc (Jet); Jan Stromme / Photonica 38c; Roger Tully 19tl; Ales Veluscek 43cr; **ICRR (Institute for Cosmic Ray Research), The University of Tokyo:** 61br; **iStockphoto.com:** 29br, 29tr, 41cr; Douglas Allen 39tr; Ivan Grishkov 2tl, 17tc; MoosyElk 44bl (Spoons); Roman Sigaev 67cra (glass pan); **The Kobal Collection:** Paramount Television 63bc; **NASA:** JPL-Caltech 18cl; NASA and G. Bacon (STScI) 60tl; NASA-HQ-GRIN. 59tl; **NAVY.mil:** Sonar Technician (Surface) 1st Class Ronald Dejarnett 30r; **PhotoEdit Inc:** Bonnie Kamin 10tr; **Photolibrary:** 31br, 53tr; Peter Arnold Images 54cl; BSIP Medical 50bl; Reinhard Dirscherl / Mauritius 27t; Peter Giovannini 22l; Javier Larrea 18-19; Hugh Morton / Superstock 39br; Paul Nevin 34-35; Oxford Scientific 13r, 57tl; Purestock 52bl; RESO 37cr; Guido Alberto Rossi / Tips Italia 24bl; **RLT.COM:** 11br; Photo Scala, **Florence:** 9br; **Science Museum / Science & Society Picture Library:** 12c, 12l, 12tr, 13l, 27cr; **Science Photo Library:** 8bl, 23br; Joel Arem 40tl; Astier-Chru Lille 63cr; A Barrington Brown 50cl; Massimo Brega, The Lighthouse 69tr; CCI Archives 11cr; Centre Jean Perrin, ISM 23crb; Martyn F Chillmaid 48bl; Thomas Deerinck NCMIR 19cr; E. R. Degginger 38bl; Equinox Graphics 60cr; Kenneth Eward 5tr, 33tr; Gustoimages 27bl; Richard R Hansen 70t; Hossler/Custom Medical Stock Photo 48br; James King-Holmes 51tc; Patrick Landmann 65br; R. Maisonneuve, Publiphoto Diffusion 31cr; Maximilian Stock Ltd 47bl; Peter Menzel 19cl; Cordelia Molloy 43tr; NASA 16-17bc, 17tr; NASA/ JHU/APL 16bl; Pasieka 48-49; Philippe Plailly 19tr; RIA Novosti 21br; Victor de Schwanberg 60bl; Science Source 15cl; Volker Steger 62-63; Sheila Terry 48cb; Joe Tucciarone 61l; US Dept of Energy 71l; Charles D Winters 3tr, 20r, 21l, 64bl; Charles D. Winters 32-33bc (dry ice); **SETI@home/University of California:** 62bl; **TopFoto.co.uk:** The Granger Collection 8tr; Dr. Elmar R. Gruber 6bc; **World of Stock:** David Ewing 36tl

Jacket: *Front:* **Corbis:** Visuals Unlimited tc; **Dorling Kindersley:** National Maritime Museum, London tr; **Science Photo Library:** James King-Holmes b; Pasieka ftr. *Back:* **Dorling Kindersley:** Royal Tyrrell Museum of Palaeontology, Alberta, Canada cr; **Getty Images:** Liu Jin/AFP tl; **Science Museum/Science & Society Picture Library:** crb; **Science Photo Library:** br; Charles D Winters cl

Wallchart: **Alamy Images**: Mesopotamian / The Art Gallery Collection (ca); sciencephotos (cb/slinky); John Warburton-Lee Photography (cra). **Corbis**: Joe McDonald (cb); NASA (bl); Bruce Benedict/ Transtock (tr/car). **Dorling Kindersley**: The Science Museum, London (cla). **Getty Images**: Colin Anderson / Brand X Pictures (c); Science Photo Library/Pasieka (cr); Brian Stablyk (truck). **iStockphoto.com**: (bc); MoosyElk (tr/spoons). **NASA**: JPL-Caltech (cl). **PhotoEdit Inc**: Bonnie Kamin (tl). **Science Museum / Science & Society Picture Library**: (tl/orrery). **Science Photo Library**: Astier-Chru Lille (br); RIA Novosti (clb/ Mendeleev); Charles D Winters (clb). **World of Stock**: David Ewing (cb)